Web Wisdom

How to Evaluate and Create Information Quality on the Web

Web Wisdom

How to Evaluate and Create
Information Quality on the Web

Janet E. Alexander
Marsha Ann Tate
Wolfgram Memorial Library
Widener University

LEA **LAWRENCE ERLBAUM ASSOCIATES, PUBLISHERS**
1999 Mahwah, New Jersey   London

Lawrence Erlbaum Associates, Inc., Publishers
10 Industrial Avenue
Mahwah, New Jersey 07430-2262

Cover design by Kathryn Houghtaling Lacey

Library of Congress Cataloging-in-Publication Data

Web wisdom : how to evaluate and create information quality on the Web / Janet E. Alexander, Marsha Ann Tate.
    p. cm.
Includes bibliographical references and index.
ISBN 0-8058-3122-3c (alk. paper) —ISBN 0-8058-3123-1p (alk. paper)
    1. Web sites. 2. World Wide Web (Information retrieval system). I. Tate, Marsha Ann. II. Title.
TK5105.888.A376 1999
005.7'2—dc21   99-12314
                                                                      CIP

Books published by Lawrence Erlbaum Associates are printed on acid-free paper, and their bindings are chosen for strength and durability

Printed in the United States of America
10 9 8 7 6 5

To my mother, Barbara, and in memory of my father, Andrew Tate, Jr., and my grandfather, Andrew Tate, Sr. Their enduring love and confidence in me made this all possible and I am forever grateful.
—Marsha Ann Tate

To my parents, Marjorie and Earle Edwards, and to my husband George, all three of whom have always provided me with encouragement and support in both my personal and professional endeavors.
—Jan Alexander

Contents

Acknowledgments

We would both like to thank the following people, each whose assistance and encouragement helped make the creation of this book possible.

At Lawrence Erlbaum Associates: Lane Akers, who from a brief query letter saw the potential for this project, and Linda Bathgate, who, after reading our proposal immediately shared our vision for the book. Linda has given us steady support and encouragement as the project developed.

At Widener University: Maria Varki, Head of the Reference Department, who has always, from the first days of the World Wide Web, stressed the need to teach students not just how to use the Web, but also how to use it well. Bob Danford, Library Director, for his continual support and inspiration; Mike Powell, for many things, not the least of which has been his knowledgeable and generous technical assistance and his help as we presented these ideas to a wide variety of audiences; Terri Cartularo, Edna Moore and our other colleagues at the Wolfgram Memorial Library who have provided continual encouragement; and Provost Larry Buck, for his ongoing support of the library and its mission

We are also indebted to Laverna Saunders, at *Computers in Libraries* magazine, and to the many individuals from the around the world who have shared their ideas about how to evaluate Web resources. Thanks also to the organizations and businesses that generously granted permission to use screen captures of their Web pages: the Ad Council, Dr. T. Matthew Ciolek, Consumers Union of U.S. and *Consumer Reports*, Lands' End Direct Merchants, the Math Forum at Swarthmore College, Minnesota Public Radio, OncoLink, Physicians for Social Responsibility, the Public Broadcasting Service, the Smithsonian Institution, the University of Pennsylvania, the Washington Post Company and washingtonpost.com, and Why? InterNetworking. We also thank the U.S. Government Environmental Protection Agency and the White House for the use of screen captures, and the Corel Corporation for the use of icons in the book's fictitious Web pages.

Acknowledgments from Marsha Ann Tate

I owe a huge debt of gratitude to my sources of inspiration and support:

- Brenda Corman, Lock Haven University. Brenda's willingness to listen to my ideas and provide her wisdom on the topic was greatly appreciated. I also would like to thank her for helping edit the manuscript and most of all for her support throughout the entire process.

- Ken Robinson, the Pennsylvania State University. Ken's wonderful whimsical sense of humor not only provided inspiration but also served as a much needed morale booster when things got rough! Thanks also to Brenda and Ken for helping create the *Neon Potato* site and for their willingness to let me use materials from their personal Web pages in the book.
- I also want to thank Barbara Coopey, Earl House, and Mary Hosterman from the Pennsylvania State University for their encouragement and support.

Acknowledgments from Jan Alexander

The members of my family made essential contributions to the writing of this book and my thanks are enormous.

- George, for his love, support, and endless patience in living with an often preoccupied author, and for his excellent editing suggestions that continually raised the quality of our work.
- Chris, who has shared his love of the Web with me from its early inception, and who appeared as just the right moments with invaluable technical assistance. Chris's knowledge of the Web, the Macintosh, and the Buddhist concept of mindfulness, and his intuition about how to get them all to work together, continually came to my rescue.
- Becky, who devoted her artistic talents to the design and creation of the annotated screen captures. Her extraordinary patience and dedication to creating quality work converted the scribblings of our ideas into the book's excellently laid out illustrations.
- Thanks also to my parents for helping me as a child to learn perseverance (an essential trait for an author) and to Dave, Barb, and Steve, whose presence makes many things go better.
- Finally, I would like to thank my mother-in-law, Mary Alice, who added a note of humor to the project by sending me the following quotation from Mark Twain: "There ain't nothing moe to write about, and I am rotten glad of it, because if I'd a knowed what a trouble it was to make a book I wouldn't a tackled it, and ain't a going to no more" (found in a genealogy of W. H. Mullin's wife's family at the Salem Historical Society Museum).

Foreword

Robert Danford
Wolfgram Memorial Library
Widener University

The enormous number and variety of Web sites today offer unparalleled amounts of information. The relative ease and economy with which individuals and institutions can offer information via the Web poses, however, quite a dilemma to the individual user of the Web: How does one find appropriate Web sites and, from those Web sites, how does one select the site or sites that are most appropriate for the immediate purpose? On the producer side of the equation, how does one ensure that the Web site under creation will present itself to users in a manner that will set it apart from other similar sites?

This work offers a model of evaluation that will assist users of the World Wide Web in selecting the best Web site for an intended purpose. Having a world of Web sites available is not necessarily a boon if one cannot judge which is the best among competing sites. Using techniques of critical thinking, the authors present tools that will help users examine all aspects of a Web site to determine its relative usefulness. Users of the Web must approach the plethora of Web sites with the skills of a good consumer to see if the product offered is indeed what it purports to be, to see if the site will fulfill the user's need, to see which of the sites available is the "best" in a given situation so that the user will be able count on the information or services offered. Just as one may act on his or her evaluation of a car or a suit of clothes, he or she may act on information found on the Web—how can we be sure that the information is accurate and useful?

Producers of Web sites also will find this book valuable. They will want to know how competing sites may be evaluated or judged. Provision of information and services on the Web is done in a highly competitive arena—how does the producer ensure that his or her site stands out among the crowd? What elements of a Web site design enhance the usefulness and authority of a site? What elements of design cheapen or weaken the authority of the site? Just as a user needs to know what elements of a Web site indicate potential value and usefulness, a producer needs to know what users will respond to as they make vital choices from the incredible variety of sites that compete for their attention.

Jan Alexander and Marsha Tate, reference librarians at Widener University's Wolfgram Memorial Library, offer a model of evaluation that uses an approach of establishing authority, accuracy, objectivity, currency, and coverage. They discuss specific problem areas that are unique to digital information and digital information presentation and that differ radically from print products.

They illustrate well the often hidden differences between types of pages (advertising pages vs. information pages vs. entertainment pages), differences that can be deciphered. Their techniques offer the consuming public a way to have the confidence needed to evaluate effectively the barrage of sites from which they must choose as they go about their work and research.

With so much information being offered via the Web, one must struggle to find the useful, the best, and the most appropriate. Employing the techniques found in this book, consumers and producers can establish an effective partnership in the use of digital information.

A Related
Web Site

A Web site created by the authors and composed of related Web evaluation materials is located at http://www.widener.edu/libraries.html (Select "Evaluating Web Resources"). The following materials are located at the site:

1. Links to many of the Web page examples used throughout the book, as well as links to numerous other sites that illustrate Web evaluation concepts.
2. A PowerPoint presentation that discusses five traditional evaluation criteria and their application to Web resources.
3. A PowerPoint presentation on the topic of advertising and sponsorship on the Web.
4. Contact information for the authors.

1

Web Wisdom:
Introduction and Overview

Chapter Contents

- **The Need for Web-Specific Evaluation Criteria**
- **What the Book Includes**
- **A Note About Design Issues**
- **How to Use the Book**
- **Two Important Caveats**
- **Definitions of Key Terms**

The World Wide Web offers us unprecedented communicative powers. It enables us to read breaking stories from news sources around the world, track population estimates on a second-by-second basis, and locate medical information on nearly every disease imaginable. In fact, the Web makes possible the instant retrieval of information on virtually any topic we care to explore. It is also revolutionizing our buying habits. We can make online plane and train reservations, and browse through countless virtual stores, purchasing merchandise from our desktops. Our unprecedented access to information and ability to communicate with others is radically transforming many aspects of our daily lives.

But how, among this extraordinary abundance of resources, do we know what to believe? How can we determine what information is authoritative, reliable, and therefore trustworthy? Although the challenge of evaluating resources is as old as information itself, the Web brings new and sometimes complicated twists to the process. This book provides tools and techniques to help meet the sometimes straightforward and sometimes convoluted evaluation challenges posed by the Web.

However, the book is not merely directed toward Web users. It also provides important guidance for Web page authors. Web authors have information they want to share with others, and they need to present this information so it can be recognized as reliable, accurate, and trustworthy. How can a Web user know whether to trust information if the Web author has not included such basic information as who is responsible for the contents of the page, and provided a way of verifying that person's credentials for writing on the topic? How can a Web user know whether to trust information if there is no viable way to determine what influence an advertiser may have on the integrity of that information? How can a Web user know whether to order products from a company if there is no way of verifying that company's legitimacy?

This book discusses these issues and describes the basic elements that all Web page authors, new or experienced, need to address. By following the suggestions in this book, there is an increased likelihood that the author's message will be more successfully conveyed to the Web user.

THE NEED FOR WEB-SPECIFIC EVALUATION CRITERIA

Today's media send out a steady stream of broadcast and print messages intended to entertain, inform, and influence the public's actions and opinions. The advent of the World Wide Web has added yet another component to this daily barrage. Based on a lifetime's exposure to media messages, we develop a set of criteria that we use to evaluate the messages received. Moreover, the same evaluative criteria that we apply to the traditional media messages can also serve as a useful starting point for developing methods for evaluating Web resources. Each individual's evaluative criteria will differ somewhat based on various demographic, social and psychological factors. However, five specific criteria—accuracy, authority, objectivity, currency and coverage—play an essential role in the evaluation process. Chapter 2 will discuss these criteria in much greater detail.

In addition, several other factors have important roles in the evaluation process. These include standards and guidelines, regulations, and our own sensory perception. Many information providers adhere to a well-established set of industry standards and conventions regarding the content and presentation of their materials. Information providers are also obliged to comply with various governmental regulations that affect the content and format of their messages.

Using visual and textual cues, an individual can usually differentiate between advertising and informational content in a magazine or newspaper. Similar distinctions occur in radio and television as well. For example, a television commercial is ordinarily distinguishable from the program itself by means of audio and visual cues. Even an infomercial, a program-length advertisement, is by law accompanied by a disclaimer proclaiming it as a "paid program."

Of course, all of these waters can, and frequently do, get muddied. Whenever a company or organization advertises in a print or broadcast medium, the potential always exists for the contents to be influenced in some manner by the advertiser. Most savvy consumers understand this situation and judge the trustworthiness of the information accordingly.

However, since the Web is such a new medium, many standards, conventions and regulations commonly found in traditional media are largely absent. Lacking many of these traditional formalities, how can a Web user evaluate the quality of Web information?

A number of resources have been developed to help Web users locate quality Web information. For example:

- Certain Web-based subject directories include qualitative reviews of sites.

- A variety of organizations provide their members with a listing of Web sites they have found valuable.
- Web pages at the academic departments of universities often have a listing of quality Web sites relevant to their subject specialties.
- Librarians create pages of authoritative links on topics of interest to their patrons.
- Magazines devote sections of an issue or an entire issue to the evaluation of Web resources.
- A number of health organizations evaluate medical-related sites.

However, as valuable as these efforts to review individual sites are, they cannot begin to cover more than a small fraction of the resources available on the Web. Moreover, although review services may purport to list sites on the basis of quality, in reality a site may be listed merely because it paid money and not because of its inherent quality. Therefore, it is still imperative that Web users know how to judge the quality of information they find on a Web page for themselves.

WHAT THIS BOOK INCLUDES

Web page evaluation strategies are introduced in chapter 2, with an overview of five traditional evaluation criteria: authority, accuracy, currency, objectivity, and coverage. Chapter 2 also discusses the more complex evaluation questions necessitated by characteristics unique to the Web—features such as the use of hypertext links and frames as well as the need for specific software to access certain materials.

Chapter 3 explores advertising and sponsorship on the Web. It addresses such issues as determining the sponsor of commercial advertising and informational content on a Web page and the possible influence an advertiser or sponsor may have on the objectivity of any information provided on the page.

Chapter 4 investigates the concepts and issues introduced in chapters 2 and 3 in greater detail, and include a checklist of basic questions to ask when evaluating or creating any type of Web page. The chapter also includes annotated screen captures of actual Web pages that illustrate many of the concepts discussed.

The next six chapters focus on analyzing different types of Web pages based on the framework established in chapters 2, 3, and 4. However, no "one size fits all" approach is adequate for analyzing the diverse array of Web pages.

Therefore, we categorize Web pages into the following six types based on the page's purpose: advocacy, business, informational, news, personal, and entertainment. For example, a business page advertising a company and its products has somewhat different goals from an advocacy page of a political party urging voters to support a legislative initiative. Likewise, a news page is significantly different from a personal page created by an individual who merely wants to share photos of the family pets. Therefore, in addition to the checklist of basic questions found in chapter 4, the book also includes checklists of additional questions to ask when evaluating or creating each specific type of Web page. Each chapter also illustrates the concepts discussed via numerous annotated screen captures in addition to page-specific checklists.

Chapter 11 essentially focuses on Web page creation issues such as:

- Consistent use of navigational aids.
- Meta tags.
- Basic copyright considerations.
- Testing the functionality of a completed Web page.

A NOTE ABOUT DESIGN ISSUES

Two important aspects of Web page design are:

- Visual design, which consists of aesthetic factors such as the use of images and color.
- Functional design, which consists of factors such as conformity of layout and use of hypertext links to aid in page navigation.

Visual design issues, although important, are well covered in other books and are not addressed here. However, functional design issues are addressed in this book since they can have a significant impact on information quality.

HOW TO USE THIS BOOK

Chapters 2, 3, and 4 are intended to be read consecutively because they serve as the conceptual foundation for the evaluation criteria and the questions that appear in checklists used throughout the book.

Chapters 5 through 10 are intended to serve as a resource for understanding the six different types of Web pages and the additional questions that need to be asked when either evaluating or creating each type of page. Consequently, these chapters can be either read through consecutively to gain an understanding of the different types of pages, or consulted individually when evaluating or creating a specific type of page.

Although chapter 11 is designed basically for those readers creating Web pages, much of the information covered, including meta tags and copyright, can be useful to Web page evaluators and Web page authors.

For the reader's convenience, a complete set of all checklists that appear throughout the book is provided in Appendix A.

To help provide continuity throughout the book, we have assigned a unique identifier, consisting of a combination of letters and numbers, to each important concept introduced in the book. The unique identifier appears each time the concept is repeated in any checklist or illustrated on a screen capture. For example, when we discuss the concept of currency, we ask the question, "Is the date the material was first placed on the server included on the page?" This question has the unique identifier CUR 1.2. All identifiers associated with the concept of currency begin with CUR. The number 1.2 following CUR refers to the specific aspect of currency discussed. In this instance, the specific aspect of currency discussed is the date the material was first placed on the server. In addition, whenever this specific concept is illustrated on a screen capture, the identifier CUR 1.2 will reappear. Each of the major concepts discussed in chapters 2 through 4 are denoted with similar letter and number combinations.

It is our hope that the unique identifiers will enable the reader to readily follow the concepts as they are explained and illustrated. Appendix B contains a complete listing of all the questions accompanied by their unique identifiers.

TWO IMPORTANT CAVEATS

This book presents a variety of methods for analyzing how Web information can be evaluated and presented. It must be noted, of course, that it is possible to follow the techniques provided in this book and construct pages and sites that outwardly appear to be trustworthy yet deliberately provide misinformation. This creates obvious problems for a Web user attempting to evaluate such pages. The Web, perhaps more than any other medium, contains these inherent dangers and, regardless of the evaluation techniques employed, there cannot be any absolute guarantees that information that seems to satisfy the criteria is always accurate and trustworthy.

We do not attempt to be arbiters of what is a "good" or "bad" Web page. In fact, without knowing the purpose for which information will be used, this judgment cannot be made. Instead, this book seeks to provide Web users with a method to help them think critically about the Web information they locate and to make their own judgments about whether the information is suitable for their needs.

As previously stated, whether this is so depends on the user's purpose for accessing the information. There may be occasions when certain criteria discussed, such as the need for indicating an author's qualifications to write about a topic, will not be important to the user. For example, if a user has sufficient expertise in a subject area to judge the information quality of a Web page directly, the page may be of value even without a listing of the author's credentials. Moreover, if someone is merely seeking fan fiction based on a favorite television show, the absence of an author's name and qualifications may not be critical.

However, in many situations, it is important to try to ascertain whether Web information is accurate, authoritative, and reliable. Because of this, we trust both Web users and Web authors will find the tools and techniques presented in this book a useful resource.

DEFINITIONS OF KEY TERMS

The Web is a new medium, with a vocabulary all its own. Because Web terminology is not always intuitively clear, and because certain key concepts are not always defined in a similar way, we want to clarify how we are using the following terms throughout the book.

- *Home page*: The page at a Web site that serves as the starting point from which other pages at the site can be accessed. A home page serves a function similar to the table of contents of a book. A home page is to be distinguished from the welcome page found at some sites that serves as a gateway to the site but does not function as the table of contents. Note that sometimes, although not in this book, the term *home page* is used to refer to an entire Web site.

- *HTML (Hypertext Markup Language):* A set of codes that are used to create a Web page. The codes control the structure and appearance of the page when it is viewed by a Web browser. They are also used to create hypertext links to other pages.
- *Hypertext link (or link.):* A region of a Web page that, once selected, causes a different Web page or a different part of the same Web page to be displayed. A link can consist of a word or phrase of text, or an image. The inclusion of links on a Web page allows users to move easily from one Web page to another.
- *Internal search engine:* A search engine that searches for words or phrases only within one World Wide Web site.
- *Search engine:* A tool that can search for words or phrases on a large number of World Wide Web pages. Examples of search engines include AltaVista, Infoseek, and HotBot.
- *Uniform resource locator (URL):* A unique identifier that distinguishes a Web page from all other World Wide Web pages.
- *Web page:* An HTML file that has a unique URL address on the World Wide Web.
- *Web site:* A collection of related Web pages interconnected by hypertext links. Each Web site usually has a home page that provides a table of contents to the rest of the pages at the site.
- *Web subsite:* A site on the World Wide Web that is nested within the larger Web site of a parent organization. The parent organization often has publishing responsibility for the subsite, and the URL for the subsite is usually based on the parent site's URL.

2

Information Quality Criteria
for Web Resources

```
┌─────────────────────────────────────────┐
│                                         │
│           Chapter Contents              │
│                                         │
│   • A Comparison Between Two            │
│     Web Pages                           │
│     Presenting Information              │
│   • Five Traditional Evaluation         │
│     Criteria and Their                  │
│     Application to Web Resources        │
│   • Additional Challenges               │
│     Presented by Web Resources          │
│                                         │
└─────────────────────────────────────────┘
```

A COMPARISON BETWEEN TWO WEB PAGES PRESENTING INFORMATION

Figures 2.1 and 2.2 are both Web pages that might be retrieved using one of the general Web search engines. Both pages have important messages to convey, yet there are striking differences in how effectively these messages are presented. Figure 2.1, a section from a Web Page entitled "Sex Differences in Children's Behavior: Genetics or Environment?," describes the results of research on sex differences in the behavior of rhesus monkeys. Although the information appears to be valid, there is no easy way to determine the information's attribution and reliability, for the following reasons:

- No author is given for the work, and there is no link to a home page that might identify the author and his or her qualifications. As a result, we have no way of knowing whether the author is a scholar in the field or a student writing a term paper.
- Without knowing the author's rationale for writing this work, we cannot adequately determine whether the material is intended to be presented in an objective manner, or whether it has been slanted by someone with a particular point of view.
- This page has become separated from the rest of the material, and there are no links to enable a reader to easily locate the other parts. As a result,

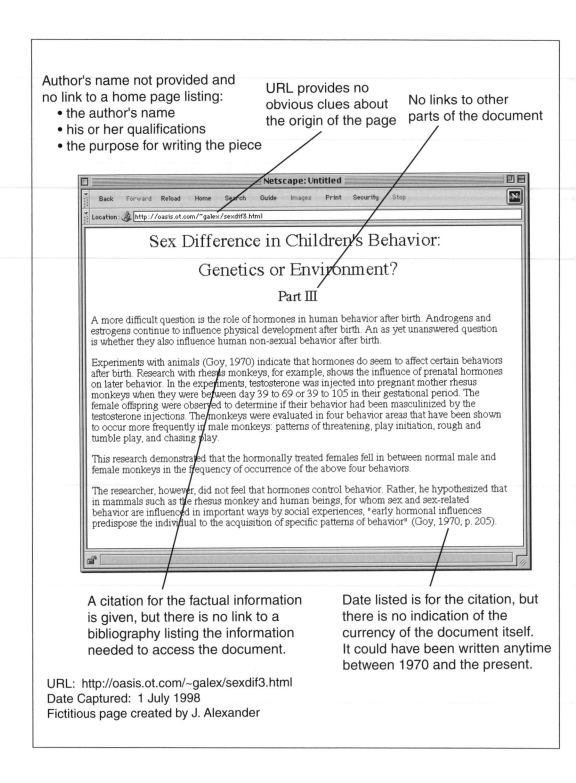

Author's name not provided and no link to a home page listing:
- the author's name
- his or her qualifications
- the purpose for writing the piece

URL provides no obvious clues about the origin of the page

No links to other parts of the document

Netscape: Untitled

Back Forward Reload Home Search Guide Images Print Security Stop

Location: http://oasis.ot.com/~galex/sexdif3.html

Sex Difference in Children's Behavior:

Genetics or Environment?

Part III

A more difficult question is the role of hormones in human behavior after birth. Androgens and estrogens continue to influence physical development after birth. An as yet unanswered question is whether they also influence human non-sexual behavior after birth.

Experiments with animals (Goy, 1970) indicate that hormones do seem to affect certain behaviors after birth. Research with rhesus monkeys, for example, shows the influence of prenatal hormones on later behavior. In the experiments, testosterone was injected into pregnant mother rhesus monkeys when they were between day 39 to 69 or 39 to 105 in their gestational period. The female offspring were observed to determine if their behavior had been masculinized by the testosterone injections. The monkeys were evaluated in four behavior areas that have been shown to occur more frequently in male monkeys: patterns of threatening, play initiation, rough and tumble play, and chasing play.

This research demonstrated that the hormonally treated females fell in between normal male and female monkeys in the frequency of occurrence of the above four behaviors.

The researcher, however, did not feel that hormones control behavior. Rather, he hypothesized that in mammals such as the rhesus monkey and human beings, for whom sex and sex-related behavior are influenced in important ways by social experiences, "early hormonal influences predispose the individual to the acquisition of specific patterns of behavior" (Goy, 1970, p. 205).

A citation for the factual information is given, but there is no link to a bibliography listing the information needed to access the document.

Date listed is for the citation, but there is no indication of the currency of the document itself. It could have been written anytime between 1970 and the present.

URL: http://oasis.ot.com/~galex/sexdif3.html
Date Captured: 1 July 1998
Fictitious page created by J. Alexander

FIG. 2.1. A Web page from a document, "Sex Differences in Children's Behavior."

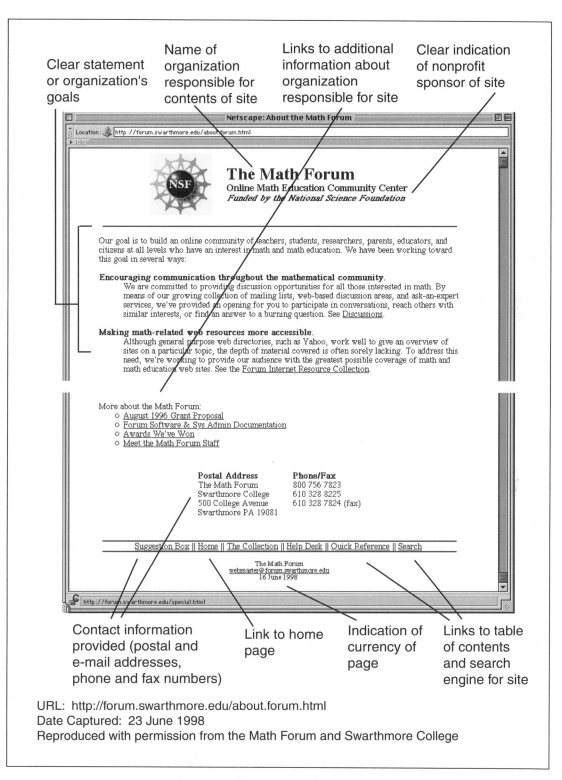

Clear statement or organization's goals

Name of organization responsible for contents of site

Links to additional information about organization responsible for site

Clear indication of nonprofit sponsor of site

Netscape: About the Math Forum

Location: http://forum.swarthmore.edu/about/forum.html

The Math Forum
Online Math Education Community Center
Funded by the National Science Foundation

Our goal is to build an online community of teachers, students, researchers, parents, educators, and citizens at all levels who have an interest in math and math education. We have been working toward this goal in several ways:

Encouraging communication throughout the mathematical community.
We are committed to providing discussion opportunities for all those interested in math. By means of our growing collection of mailing lists, web-based discussion areas, and ask-an-expert services, we've provided an opening for you to participate in conversations, reach others with similar interests, or find an answer to a burning question. See Discussions.

Making math-related web resources more accessible.
Although general purpose web directories, such as Yahoo, work well to give an overview of sites on a particular topic, the depth of material covered is often sorely lacking. To address this need, we're working to provide our audience with the greatest possible coverage of math and math education web sites. See the Forum Internet Resource Collection.

More about the Math Forum:
- August 1996 Grant Proposal
- Forum Software & Sys Admin Documentation
- Awards We've Won
- Meet the Math Forum Staff

Postal Address
The Math Forum
Swarthmore College
500 College Avenue
Swarthmore PA 19081

Phone/Fax
800 756 7823
610 328 8225
610 328 7824 (fax)

Suggestion Box || Home || The Collection || Help Desk || Quick Reference || Search

The Math Forum
webmaster@forum.swarthmore.edu
16 June 1998

http://forum.swarthmore.edu/special.html

Contact information provided (postal and e-mail addresses, phone and fax numbers)

Link to home page

Indication of currency of page

Links to table of contents and search engine for site

URL: http://forum.swarthmore.edu/about.forum.html
Date Captured: 23 June 1998
Reproduced with permission from the Math Forum and Swarthmore College

FIG. 2.2. A Web page from the Math Forum.

we cannot determine what other topics are included in the work, and to what depth these topics are addressed.

- A brief citation is provided for the factual information included in the writing. However, since the page has become separated from its bibliography, we cannot access the full citation, which would be needed to retrieve the original work and validate the facts presented.
- The only date on the page, 1970, is found in the citation. Thus, the original document could have been written anytime between 1970 and the present.

Although the information on this Web page may indeed be from a reliable source, we have no assurance that this is so because of the way the information is presented on the Web.

In contrast, Fig. 2.2, a page from the Math Forum at Swarthmore College, provides us with the following information that we can use to help determine its authorship and reliability:

- The page clearly indicates what organization is responsible for the information.
- Numerous contact points are listed that can be used to verify the legitimacy of the organization.
- The purpose of the site is described.
- There is a link to the home page of the organization responsible for the page.
- A date on the page indicates the currency of the information.
- A link is provided to the site's table of contents to enable a user to learn more about what is included at the site.
- The site's nonprofit sponsor is clearly indicated.

Although Web users may never had heard of the Math Forum before, the page provides enough evidence to help them determine whether the information on it is likely to be trustworthy.

Both of these pages convey what appears to be valuable information, yet there is a great disparity between them with respect to verifying the quality of the information provided. This chapter discusses the factors that must be addressed in order to present information that can be identified as reliable and authoritative. Understanding these same factors will also aid Web users in determining whether the information they retrieve is coming from reliable, trustworthy sources.

FIVE TRADITIONAL EVALUATION CRITERIA AND THEIR APPLICATION TO WEB RESOURCES

We include here a discussion of five traditional evaluation criteria—authority, accuracy, objectivity, currency, and coverage. These criteria have their origins in the world of print, but are universal criteria that need to be addressed regardless of the medium being evaluated. To provide a more in-depth understanding of each criterion, we address each one individually. However, we also discuss the significant overlap that can occur between criteria. For example, authority and accuracy often go hand in hand and may need to be viewed together to get a more complete picture of a particular resource.

1. Authority

Authority is the extent to which material is the creation of a person or organization that is recognized as having definitive knowledge of a given subject area.

Authority of Traditional Sources.
There are several methods to assess the authority of a work. One way is to determine an author's qualifications for writing on the subject by examining his or her background, experience, and formal credentials related to the subject area.

Another method for assessing the authority of a work is to examine the publisher's reputation. A publisher earns a reputation for the quality of its materials based on numerous factors, such as:

- The accuracy of the content in its products.
- The types of people who use the products.
- Reviews written about the publisher's products.
- The expertise of the authors writing for the publisher.

A publisher that wants to produce quality works must establish and adhere to strict editorial and ethical guidelines that emphasize quality. The publisher employs editors and ombudsmen who continually monitor the information presented. If these practices are consistently and effectively used the publisher should gain a reputation for producing publications of excellence and integrity. For example, the publisher of the *Encyclopedia Britannica* has gained a reputation for producing high-quality work largely by employing this process.

Authority of Web Sources.
One of the factors that has contributed to the explosive popularity of the Web is the ease with which almost anyone can become a Web publisher. Countless people can now easily circumvent the traditional publishing process and share their message directly with a worldwide audience. This factor, although one of the Web's great strengths also poses new evaluation challenges.

On the Web, the search for clues with which to adequately evaluate a work can be quite difficult. For example, as demonstrated in Fig. 2.1, there is no guarantee an author's name will be given, or that his or her qualifications will be provided. Also, if an author's name is given on a page, it should not be automatically assumed that this person is the actual author. In addition, it is often difficult to verify who, if anyone, has ultimate responsibility for publishing the material.

2. Accuracy

Accuracy is the extent to which information is reliable and free from errors.

Accuracy of Traditional Sources.
Traditional media utilizes a number of checks and balances to help assure the accuracy of content. These include:

- The use of editors and fact checkers to monitor accuracy.
- The peer review process to monitor the accuracy of scholarly journal articles.
- The use of style manuals to promote uniformity in language usage and manuscript format.
- The listing of sources for factual information, as appropriate.

Evaluating information encompasses a large part of our daily lives, yet we often are not consciously aware of the process. Even a simple trip to the supermarket requires making a large array of evaluation decisions. We often compare products on the basis of such objective and subjective criteria as ingredients, price, size, and even shelf location and the package look. Frequently, our past experience with a particular brand name plays a large role in our purchasing decisions. For example, if we purchased XYZ brand spaghetti sauce in the past and found it to be of high quality, the XYZ brand has earned a good reputation in our view. As a result, we will probably be more likely to purchase XYZ's spaghetti sauce in the future and also would be more apt to buy another XYZ brand product if making a decision between XYZ and an unfamiliar brand name.

We even evaluate information while we wait in the checkout line and scan through the tabloids. Once again, reputation plays a role in our evaluation of these tabloid stories. However, in this instance, our focus is on the publisher's reputation for accuracy, objectivity, and so forth. As a result, we tend to give more credence to information found in *The New York Times* rather than the information offered in a tabloid on a related subject. As the above example demonstrates, reputation often influences our differentiation between the quality of products, news, and many other items. Consequently, we revisit the reputation factor several times throughout this book.

As mentioned earlier, authority and accuracy are often interrelated. We often make the assumption, usually with good reason, that a publisher with a reputation for reliability will produce works that are also accurate. The *World Book Encyclopedia*, for example, is a publication found in countless schools and homes because it has a reputation as an authoritative, reliable publication. Although users may not know that the publisher of the *World Book* has a policy whereby the facts presented are verified in three different primary sources, they do assume, because of the publisher's reputation, that information found in it will be accurate.

Accuracy of Web Sources.   As stated previously, one of the benefits of the Web is that people can easily make their works public, independent of the traditional publishing or broadcasting process. Another major advantage of the Web is its timeliness, as Web material can be published almost instantaneously. However, the steps that contribute to the accuracy of traditional media are frequently condensed or even eliminated on the Web.

This condensation of the traditional publishing process can result in problems as straightforward as the omission of a listing of sources used in research, or as complex as what happened to a major news provider in June 1998, when it prematurely published the obituary of Bob Hope, who was very much alive at the time (Stout, 1998). In the case of the Bob Hope obituary, the source of the information was authoritative, but the Web publishing process had somehow circumvented the organization's checks and balances usually in place to assure accuracy.

3. Objectivity

Objectivity is the extent to which material expresses facts or information without distortion by personal feelings or other biases.

Objectivity of Traditional Sources.
No presentation of information can ever be considered totally free of bias, because everyone has a motive for conveying a message. However, it is often important to try to assess the information provider's objectivity. Knowing the intent of the organization or person for providing the information can shed light on any biases that might be present in the material. For example, we would easily be able to evaluate the objectivity of information originating from the Surgeon General or the tobacco lobby. However, for information sources we are not familiar with, unless the material states its point of view, it can be very difficult, even in print sources, to evaluate the objectivity of its contents.

Objectivity of Web Sources.
If we are familiar with the source of information on the Web, evaluating its objectivity is probably no more difficult than evaluating the objectivity of print information. However, because the Web so easily offers the opportunity for persons or groups of any size to present their point of view, it often functions as a virtual soapbox. It can be difficult, in this jumble of virtual soapboxes, to determine the objectivity of many Web pages unless the purpose of the individual or group presenting the information is clearly stated.

When discussing objectivity, another important factor to consider is the potential influence exerted by advertisers or sponsors on the informational content of materials. Although the extent of this influence is difficult to ascertain even in non-Web sources, it has become yet more complex on the Web. Because of its complexity, this issue is discussed in much greater detail in chapter 3.

4. Currency

Currency is the extent to which material can be identified as up to date.

Currency of Traditional Sources.
To evaluate the currency of a print source, it is important to know when the material was first published. For print sources, this information can usually be determined from the publication and copyright dates. However, specific kinds of material may also require additional date-related information beyond just the publication date. For example, for statistical information, it is important to know not just the publication date, but also the date the original statistics were compiled. The publication date for the *Statistical Abstract of the United States* may be 1998, but a closer analysis of the contents may reveal that the information in many of the charts was collected several years prior to publication.

In conventional print publications, the publication and copyright dates are readily available. In addition, for reputable print publications that present statistical information, the date the statistics were collected is also usually indicated.

Currency of Web Sources. Because there are no established guidelines for including dates on Web pages, it can be difficult to determine the currency of Web resources. Frequently, dates of publication are not included on Web pages, and if included, a date may be variously interpreted as the date when the material was first created, when it was placed on the Web, or when the Web page was last revised.

An advantage of Web publishing is the ease with which material can be revised. However, unless each revision is clearly dated it can be difficult to keep track of the various versions. This is obviously important if a printed or electronic copy has been made of the page for use in research and the Web page must be cited correctly, but it is also important if the page must be referred to later for any reason. In addition, because there is no standard format for how dates appear on Web pages, Web users may construe dates differently. Confusion can result when people use different conventions to convey the same information.

5. Coverage and Intended Audience

Coverage is the range of topics included in a work and the depth to which those topics are addressed. *Intended audience* is the group of people for whom material was created.

Coverage and Intended Audience of Traditional Sources. Print sources frequently include a preface or introduction at the beginning of the publication explaining the topics the work includes, the depth or level to which these topics are addressed, and the intended audience for the material. If this explanatory material is not included, a table of contents or an index may provide similar information. Even if lacking all of these features, a print source can usually be scanned or browsed through to determine the coverage of information and the audience for whom it is written.

Coverage and Intended Audience of Web Sources. On the Web, because sites often lack the Web equivalent to a preface or introduction, the coverage and intended audience of the material is often not readily apparent. Because "thumbing" through Web pages is often a tedious process, it is usually only if a site includes an index or site map that the range of topics and the depth to which they are covered can be readily determined.

ADDITIONAL CHALLENGES PRESENTED BY WEB RESOURCES

The Web is essentially a hybrid medium that integrates many components of traditional media. Like print, it has the ability to combine visual content with text. Like film and television, the Web is also capable of combining sound and video content. Another component, hypertext links, has been added to this media mix. This hypertext capability facilitates user interaction with the medium by allowing users to make choices concerning how and in what sequence they access material.

This merging of text, images, motion, sound, and interactive links constitutes a powerful new medium for conveying messages, and this hybrid medium can at times pose some difficult evaluation challenges. Two of the evaluation challenges of this new medium relate to advertising—the blending of information and advertising, and the blending of information, advertising, and entertainment. Although these advertising-related challenges are also prevalent in print and broadcast media, on the Web they can be even more difficult to evaluate than in more traditional media. Chapter 3 is devoted to these issues.

Some of the other difficult evaluation challenges posed by the Web are not found in traditional media. These challenges, unique to the Web, include:

- The use of hypertext links.
- The use of frames.
- Search engines that retrieve pages out of context.
- Software requirements that limit access to information.
- The instability of Web pages.
- The susceptibility of Web pages to alteration.

The Use of Hypertext Links

The ability to use hypertext to link a variety of pages together is one of the Web's most appealing features. However, the fact that one Web page contains material of high information quality does not guarantee that pages linked to the original page will be uniform in quality. As a result, each Web page must be evaluated independently for the quality of the information it contains.

The Use of Frames

Information presented on Web pages within frames can also present an evaluation challenge. A *frame* is a Web feature that allows the division of a user's browser window into several regions, each of which contains a different HTML page. The boundaries between frames may be visible or invisible. Sometimes each frame can be changed individually, and sometimes one frame in the browser window remains constant while the other frames can be changed by the user.

Most of the time the contents of the various frames originate from the same site. However, it is possible for the different frames to originate from different sites without the user being aware of it. A Web user needs to be alert to the fact that, because the contents of each frame may be originating from a different Web site, each frame needs to be evaluated independently.

Pages Retrieved Out of Context by Search Engines

Most Web sites are designed with the expectation that a user will initially view a page containing background information such as that provided on a home page.

However, sometimes users first enter the site at another page, as when they retrieve a page by using a search engine. In this situation, there may be no way to determine who is responsible for the page unless this information is provided either on the page itself or on a page linked to it. We encountered an example of this situation earlier in the chapter, in the illustration of the document "Sex Differences in Children's Behavior," which did not provide a link to a home page or any other identifying information. Although it is not always possible to evaluate the authority of such a page, some techniques that can help will be outlined in chapter 4.

Software Requirements That Limit Access to Information

Two factors that may limit the ability of a user to access all the information on a page are the type of browser being used and additional software that may be required to utilize the material.

Different browsers display information in varying ways. As a result, material written to be viewed by one graphical browser may not appear in the same manner when it is viewed by a different one. Similarly, material written only for a graphical browser may be completely unreadable by someone using a text-based one.

With increasing frequency, pages are being written that require additional software or hardware to access their full contents. Pages may require a sound card and the appropriate software plug-ins to access the information, or a computer may have to have a specific software module installed. When evaluating the contents of a page, it is important to realize that because of these limitations, the full contents of the page may not be accessible.

The Susceptibility of Web Pages to Alteration

Web pages are susceptible to alteration, both accidental and deliberate. For example, accidental alteration can occur when transferring information to the Web. A table that appears correctly in print may be illegible if it is converted incorrectly into another format for Web viewing. Also, technical problems with the transmission of data across the Web can cause odd characters to appear on the page or prevent the entire page from loading.

In addition, deliberate alteration can result when hackers break into a site and deliberately change the information. Because of the susceptibility of Web information to alteration, it is always important to compare information found on the Web with that found in other Web and non-Web sources to verify its accuracy.

Instability of Web Pages

The Web is inherently a less stable medium than print. Because of this, there is no guarantee that information on the Web will continue to be available in the future. Sites come and go, and the addresses of some sites and pages frequently change. As a result, the contents of a page may change or the page itself may no longer be available when a user attempts to revisit it. Unfortunately, there is little Web users can do about this situation except to be aware of it and, when using the Web for research, document Web sources by writing down their addresses and making electronic or print copies of important pages.

Web page authors can take steps to help minimize the difficulties created by the volatility of the Web. These techniques will be addressed in later chapters of this book.

Advertising and Sponsorship on the Web

Chapter Contents

- **Advertising, Sponsorship, and Information on the Web**
- **Defining Advertising and Sponsorship**
- **Distinguishing Advertising, Sponsorship, and Information on the Web**
- **Sorting Out Relations Between Advertisers, Sponsors, and Information**

ADVERTISING, SPONSORSHIP, AND INFORMATION ON THE WEB

Advertising and sponsorship are hardly new phenomena in American culture. They have long been the mainstays of newspapers and television shows, as well as art, music, sporting, and countless other activities. Advertising and sponsorship have traditionally served as a means for businesses and organizations to promote their products, services, and ideas in return for contributing financial and other support for activities.

However, the Web has introduced several new twists to traditional advertising and sponsorship. The multimedia nature of the Web, in addition to innovations such as hypertext links, frames, and cookies, has encouraged the formation of a wide array of alliances among advertisers, sponsors, and information providers.

As a result, Web users often face a daunting task when trying to ascertain the influence an advertiser or sponsor may exert on the objectivity of information provided at a Web site.

From the Web user's perspective, the depth of analysis needed of these potential influences largely depends on how he or she ultimately uses the acquired information. For example, it would certainly be more critical to determine the potential influence of an advertiser when seeking medical information or shopping for a new car than when looking for advice on buying a new compact disc. Therefore, in some cases it may be more important than in others to untangle these relationships.

DEFINING ADVERTISING AND SPONSORSHIP

Because advertising and sponsorship play significant roles in our everyday lives, we assumed it would be an easy task to find universally accepted definitions for the terms. Unfortunately, we discovered otherwise. It seems scholars, businesspeople, marketers, and the general public each ascribe somewhat different meanings to the terms. In some instances, advertising and sponsorship are treated distinctly, whereas in other instances the terms seem almost interchangeable.

However, for the purposes of this book, we define *advertising* as the conveyance of persuasive information, frequently by paid announcements and other notices, about products, services, or ideas. We define *sponsorship* as financial or other support given by an individual, business, or organization for something, usually in return for some form of public recognition.

Because these definitions encompass a diverse array of activities, we have subdivided them into the following categories.

Commercial Advertising

Commercial advertising is "advertising that involves commercial interests rather than advocating a social or political cause" (Richards, 1995–1996; http://advertising.utexas.edu/research/terms/index.html). It is designed to sell a specific product or service. Usually, the consumer can readily identify the product or service being sold. Commercial advertising can appear in a number of guises:

- Ads in print newspapers and magazines.
- Radio and television commercials.
- Billboards.
- Product placement, the visual or verbal reference to a product in another form of communication. For example, companies often pay studios a fee to have their products appear or be mentioned by a character in a movie or television show.
- Endorsements and testimonials.
- Direct mail brochures.
- Web banner ads.
- Web sites owned by a company designed primarily to promote that company's products and services.

Figure 3.1 illustrates two common forms of commercial advertising on the Web, a banner ad and the home page from a company Web site that includes promotional materials for the company's products.

Example of a Web banner advertisement from a page at the washingtonpost.com Web site

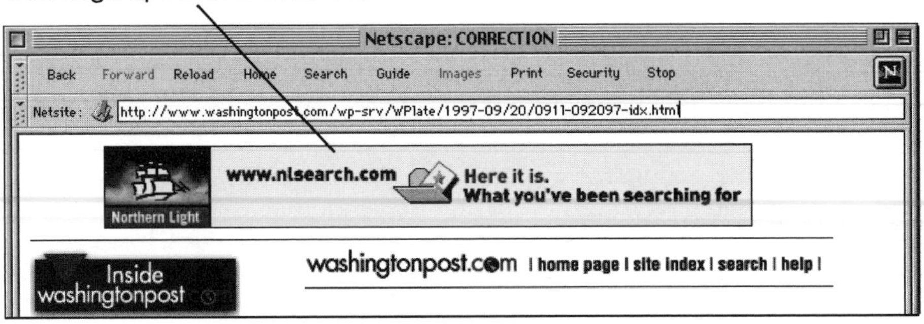

URL: http://www.washingtonpost.com/wp-srv/WPlate/1997-09/20/0911-092097-idx.html
Date captured: 11 February 1998
Reproduced with permission from washingtonpost.com

Example of the home page of a Web site owned by Lands' End Direct Merchants and designed to promote their products

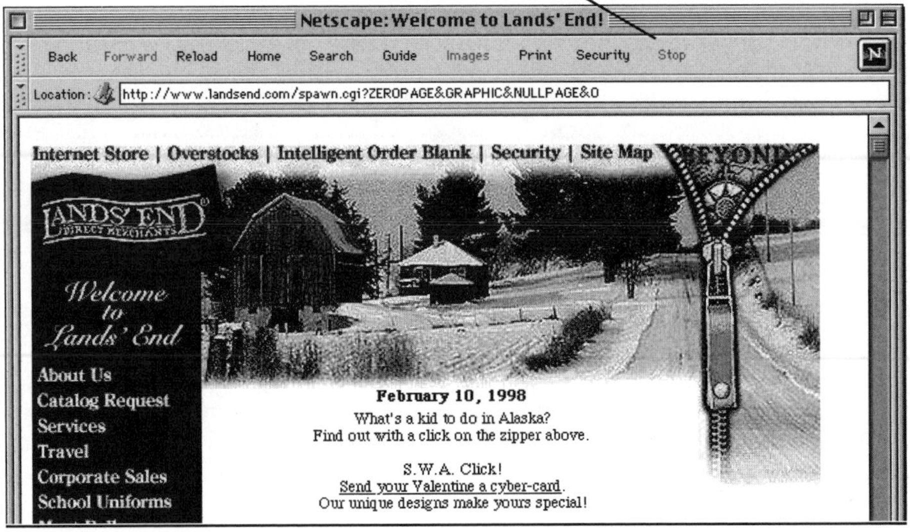

URL: http://www.landsend.com
Date captured: 10 February 1998
Reproduced with permission from Lands' End Direct Merchants

FIG. 3.1. Two examples of commercial advertising.

Advocacy Advertising

Advocacy advertising is "advertising used to promote a position on a political, controversial or other social issue" (Richards, 1995–1996: http://advertising.utexas.edu/research/terms/index.html). Examples of advocacy advertising include ads to promote:

* Public health, such as youth antismoking and AIDS prevention.
* Public safety, such as fire prevention and the use of seat belts.
* Education-related issues.

Government agencies and nonprofit organizations are often sources for advocacy advertising. Figure 3.2 provides examples of advocacy Web banner ads created by the Ad Council, whose stated mission is "to identify a select number of significant public issues and stimulate action on those issues through communications programs that make a measurable difference in our society" (Ad Council, 1997; http://www.adcouncil.org/about.html).

Institutional Advertising

Institutional advertising is "advertising to promote an institution or organization rather than a product or service, in order to create public support and goodwill" (Richards, 1995–1996; http://advertising.utexas.edu/research/terms/index.html). Institutional advertising is meant to convey the idea that the organization enhances the community in some way.

Figure 3.3 illustrates institutional advertising. In this instance, Giant Foods is promoting the "Apples for Students" program. Giant donates money toward computer-related equipment for participating schools based on the amount of money families in the schools have spent on groceries purchased at Giant.

Word of Mouth Advertising

Word of mouth advertising is the endorsement of a product or service by an individual who has no affiliation with that product or service other than being a user of it, and who is not being paid for the endorsement. Examples of word of mouth advertising would include recommending a product or service to a friend during a conversation or an individual mentioning a product on his or her Web page or in a newsgroup posting.

Corporate Sponsorship

Corporate sponsorship occurs when a company provides financial or other material support for something, usually in return for some form of public recognition. Sporting events, a wide range of artistic and educational activities, and even Web sites are often either partially or fully supported through corporate sponsorship.

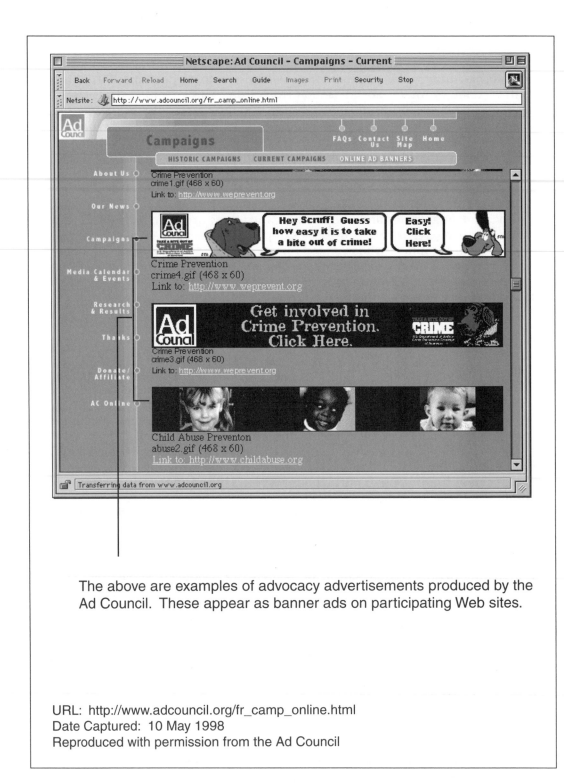

The above are examples of advocacy advertisements produced by the Ad Council. These appear as banner ads on participating Web sites.

URL: http://www.adcouncil.org/fr_camp_online.html
Date Captured: 10 May 1998
Reproduced with permission from the Ad Council

FIG. 3.2. Example of advocacy advertising.

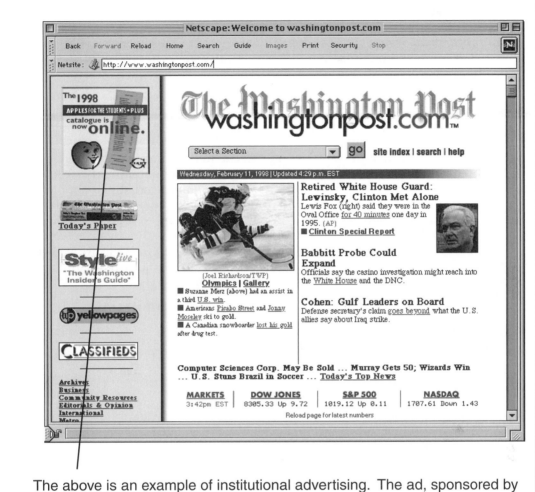

The above is an example of institutional advertising. The ad, sponsored by Giant Food, appears as a banner ad on the home page of the Web site washingtonpost.com. The company is advertising that it will donate money toward equipment for schools in exchange for cash register receipts.

URL: http://www.washingtonpost.com
Date Captured: 11 February 1998
Reproduced with permission from washingtonpost.com

FIG. 3.3. Example of institutional advertising.

Nonprofit Sponsorship

Nonprofit sponsorship consists of financial or other material support by an individual or nonprofit organization for something, usually in return for public recognition.

Although the categorization of advertising and sponsorship is a beneficial theoretical exercise, in reality, the types often defy such orderly compartmentalization. On the contrary, activities are frequently sponsored jointly by commercial and nonprofit entities. The Math Forum Web site displayed in Fig. 3.4 illustrates an example of joint commercial and nonprofit sponsorship.

Moreover, different types of advertising and sponsorship are frequently blended together so extensively that they become almost indistinguishable. This blending commonly occurs with institutional and commercial advertising. For example, company sponsorship of a sporting event can help promote the company's image (institutional advertising) with the fans. In addition, advertisements for the company's products can be readily incorporated into the event through athletes wearing the company's logo on their clothing or via banners, billboards, and countless other ways (commercial advertising).

DISTINGUISHING ADVERTISING, SPONSORSHIP, AND INFORMATION ON THE WEB

Overlapping and Blending of Advertising and Sponsorship on the Web

The concepts of advertising and sponsorship frequently overlap when they are incorporated into Web sites. Dr. T. Matthew Ciolek, an Australian University professor who maintains seveal different types of Web sites, made the following comparison between advertising and sponsorship on the Web. He described a *sponsor* as a person, company, or organization that recognizes the inherent worth and quality of a Web site and provides financial or other support toward the upkeep of the site. In contrast, an *advertiser* recognizes business opportunities offered by a Web site, and determines it is cost-effective to place an advertisement on that site. The advertiser then purchases a certain amount of exposure on the page. There is usually an agreed-on amount of exposure determined by the square inches of space, the location on the page, and the period of time for which the ad will appear (Ciolek, personal communication, March 27, 1998).

Although the difference between advertising and sponsorship is clear to the Web site owner, it is not always obvious to the user. This is partly due to the fact that both advertisers and sponsors are often identified via a banner ad on a Web page. Without a clear explanation of whether the banner ad signifies advertising or sponsorship, it is difficult for the user to differentiate between the two. In addition, the banner ads of both advertisers and sponsors are frequently linked to the Web site of the advertiser or sponsor. Thus, when a site has a corporate sponsor, it often includes a link from its home page to the site of the corporate sponsor. These links can facilitate an immediate transition from the announcement of corporate sponsorship appearing on a Web site to commercial advertising offered at the corporate sponsor's own Web site. Thus, for the user, there may be little or no apparent differentiation between the announcement of an advertiser and the announcement of a corporate sponsor.

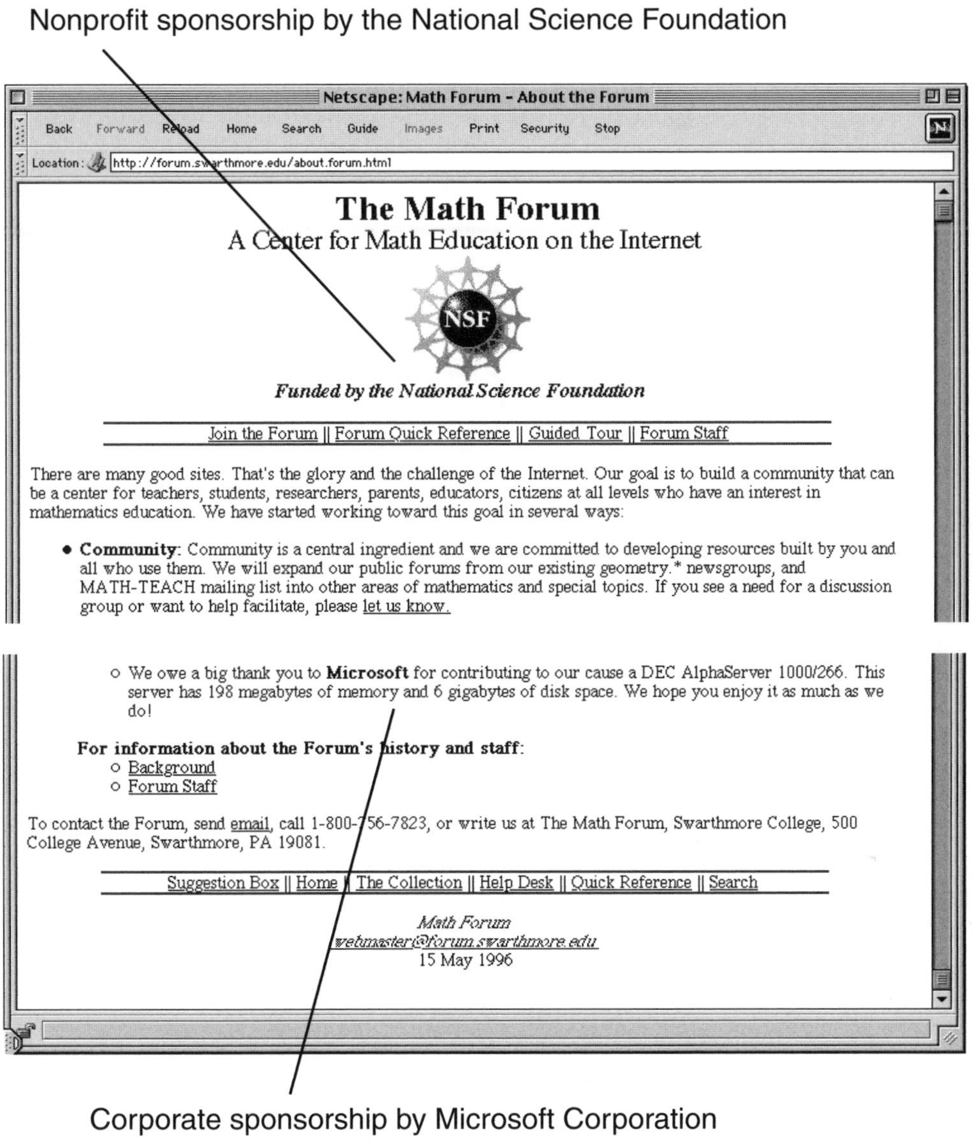

Nonprofit sponsorship by the National Science Foundation

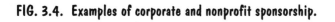

Netscape: Math Forum – About the Forum

Back | Forward | Reload | Home | Search | Guide | Images | Print | Security | Stop

Location: http://forum.swarthmore.edu/about.forum.html

The Math Forum
A Center for Math Education on the Internet

Funded by the National Science Foundation

Join the Forum || Forum Quick Reference || Guided Tour || Forum Staff

There are many good sites. That's the glory and the challenge of the Internet. Our goal is to build a community that can be a center for teachers, students, researchers, parents, educators, citizens at all levels who have an interest in mathematics education. We have started working toward this goal in several ways:

- **Community**: Community is a central ingredient and we are committed to developing resources built by you and all who use them. We will expand our public forums from our existing geometry.* newsgroups, and MATH-TEACH mailing list into other areas of mathematics and special topics. If you see a need for a discussion group or want to help facilitate, please let us know.

- We owe a big thank you to **Microsoft** for contributing to our cause a DEC AlphaServer 1000/266. This server has 198 megabytes of memory and 6 gigabytes of disk space. We hope you enjoy it as much as we do!

For information about the Forum's history and staff:
- Background
- Forum Staff

To contact the Forum, send email, call 1-800-756-7823, or write us at The Math Forum, Swarthmore College, 500 College Avenue, Swarthmore, PA 19081.

Suggestion Box || Home || The Collection || Help Desk || Quick Reference || Search

Math Forum
webmaster@forum.swarthmore.edu
15 May 1996

Corporate sponsorship by Microsoft Corporation

URL: http:/forum.swarthmore.edu/about.forum.html
Date captured: 28 March 1998
Reproduced with permission from the Math Forum and Swarthmore College

FIG. 3.4. Examples of corporate and nonprofit sponsorship.

The Blending of Advertising and Information on the Web

Just as there is often a blending of advertising and sponsorship on the Web, there can also be a blending of advertising and information on the Web. For example, a business Web site often not only promotes a product or service but may also provide a significant amount of additional apparently objective information to entice a potential customer to the site.

Some medical sites, for example, are sponsored by a doctor who provides seemingly objective information about a specific medical problem. However, the same doctor may also be promoting his or her services in the form of a surgical procedure to solve that malady. In this case, there is a definite conflict of interest and the information provided must be viewed accordingly. The critical thing to remember is that, although Web resources often provide valuable free information, the user must always ask what potential factors may influence the objectivity and thus the trustworthiness of the information.

To more fully understand the complex relations between sponsorship, advertising, and information on the Web, it is useful to explore ways sponsorship and advertising have interacted with informational content in print publishing.

In traditional print publishing, there are usually clear visual distinctions between the advertising and the editorial, or informational, content of the page. Even when advertising and information are mixed, as in the case of an advertorial that presents a significant amount of information but is in reality an advertisement for something, the print convention is to identify the information as an advertisement somewhere on the page. Thus, the phrase "Special Advertising Feature" (or something similar) often appears on the page to alert readers that what follows is information carefully blended with an advertisement. Smart consumers know to beware of the objectivity of information presented in this manner.

However, on the Web there are few, if any, standards to ensure that a visual distinction exists between advertising and information, or that advertorial material be labeled as such. As a result, information on the Web is often seamlessly blended with advertising.

In print publishing there also exists the concept of the *Chinese Wall*, a policy of separation between the advertising and editorial departments (i.e., the department that produces the information content). Where the Chinese Wall policy is more rigorously adhered to, the advertising department is not supposed to influence the editorial department. The goal is to have the information produced by the editorial department be as free as possible from advertisers' pressure to bias the information in some matter. In reality, of course, there is great variation in how strongly the concept of the Chinese Wall is adhered to by various publications.

As stated earlier, the Web is essentially devoid of any established conventions or anything that remotely resembles a Chinese Wall. One notable exception to this is the American Society of Magazine Editors (ASME), which has addressed this issue in its publication "The ASME Guidelines for New Media" (ASME, 1997; http://webreview.com/97/10/03/feature/guide/html). These guidelines state:

> The same ASME principles that mandate distinct treatment of editorial content, advertisements, and special advertising sections ("advertorials") in print publications also apply to electronic editorial products bearing the names of print magazines or offering themselves as electronic magazines....Therefore, it is the responsibility of each online publication to make clear to its users which online content is editorial

and which is advertising and to prevent any juxtaposition that gives the impression that editorial material was created for—or influenced by—advertisers.

ASME's effort to preserve the Chinese Wall is a rarity on the Web however. Thus, the Web user needs to be constantly vigilant of when advertisers might be influencing the objectivity of information.

The Chinese Wall concept is not even applicable, of course, to a large portion of the traditional media. For example, in many print publications the advertising and the informational content are produced by the same organization, as is the case with promotional brochures such as those put out by a business to advertise its trade. However, in the world of print publishing, readers have learned to recognize most publications of this type. We do not assume that a brochure produced by a car dealer is going to provide objective information about the company's product, and we know how to evaluate the material accordingly.

Similarly, for many Web sites with advertising on them, the advertiser and the organization with overall responsibility for the site are the same entity. Therefore, the material at these Web sites has more in common with a car dealer's brochure than with the information in a magazine with separate advertising and editorial departments. However, on the Web, it is often not so readily apparent when an individual or group is supplying both the informational and advertising content of the page. In addition, users of the Web information do not have the same years of experience analyzing Web resources that they do for print resources.

A Continuum of Objectivity on the Web

To better understand the potential effects of the blending of advertising, sponsorship, and information on the Web, it is helpful to view Web sites on a continuum, from sites that accept no outside advertising to sites essentially totally composed of advertising. At one end of the continuum is a site such as the *Consumer Reports* Web site. The site, like the printed magazine, accepts no outside advertising to remain free of any influence from manufacturers, either real or perceived, in its product testing. Figure 3.5 illustrates the Consumers Union No-Commercialization Policy.

At the opposite end of the continuum from the *Consumer Reports* Web site would be a site designed exclusively for marketing a company's own products and services. It is likely that any information provided on a site such a this would be strongly influenced by the company responsible for the site. Many Web sites lie between the two extremes of this continuum. Various forms of sponsorship and advertising are incorporated into these sites and the amount of their influence on the objectively of the information provided varies as well. Nonetheless, whenever any site accepts advertising and sponsorship but also provides information, the user must be aware of the potential influence by the advertiser or sponsor on the objectivity of that information.

Hypertext Links and the Blending of Advertising, Information, and Entertainment

Hypertext links play a prominent role in the blending of advertising, information, and entertainment content on the Web.

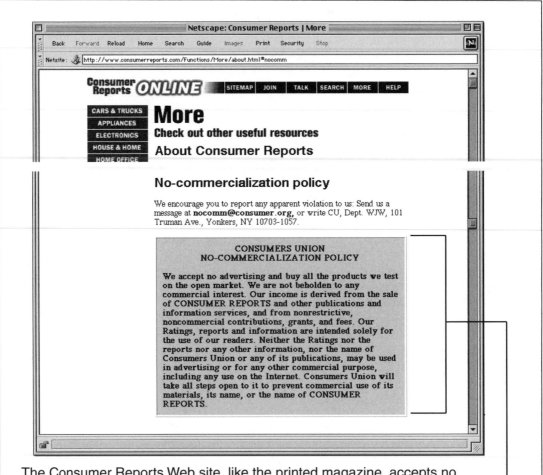

The Consumer Reports Web site, like the printed magazine, accepts no outside advertising in order to remain free of any bias, either real or perceived, in its product testing. The site clearly states the Consumers Union No-Commercialization Policy:

> We accept no advertising and buy all the products we test on the open market. We are not beholden to any commercial interests. Our income is derived from the sale of CONSUMER REPORTS and publications and information services, and from nonrestrictive, noncommercial grants, and fees.

URL: http://www.consumerreports.com/Functions/More/about.html#nocomm
Date Captured: 7 March 1998
Copyright 1998 by Consumers Union of U.S., Inc., Yonkers, NY 10703-1057.
Reproduced by permission from CONSUMER REPORTS, March 1998

FIG. 3.5. Example of a site with no outside advertising.

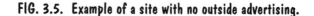

For example:

- Outside advertisers are highly motivated by marketing concerns to place a link on a site that, once followed, attracts customers from the original site and into their own site.
- Some business sites lure people to the company-sponsored site by providing links to free entertainment. Once at the site, these people can be readily marketed to.
- As a way of attracting people, business sites also may offer a listing of links to information perceived to be objective. Once again, the intention is that these visitors will respond to the company's advertising while using the site.

SORTING OUT RELATIONS BETWEEN ADVERTISERS, SPONSORS, AND INFORMATION

Because interrelations between advertising, sponsorship, and information on the Web are often more labyrinthine than those in traditional media, Web users need to become adept at sorting out these interrelations. They must learn to identify the key players and analyze as best as possible what influence they might have on each other and on the objectivity of the information provided. The following are some useful questions to ask about advertisers and sponsors found on a Web site's pages.

If the information on the page is being provided for free:

- What seems to be the purpose of the information provider for making the information available? Is the purpose one that might influence the objectivity of the information?
- What are some of the possible influences on the objectivity of the information at the site? For example, what are potential influences from the underlying motives of commercial advertisers, corporate sponsors, or the author of the information?

In addition to knowing whether the information on the page is provided for free, it is also important to know what kind of organization is providing the information. Some of the types of organizations providing information on the Web include the following.

Advocacy Groups.   When an advocacy group offers information on the Web for free, users should assume the information will be biased in a certain direction to support the organization's goals. Even if the organization provides information from a reputable journal or other outside source, users can assume they will not find both sides of the issue represented.

Nonprofit Organizations.   Even when information appears to be provided by a source such as a nonprofit organization, users must be aware of potential conflicts of interest that might arise. For example, when a piece of research is presented on a nonprofit hospital's Web site, a corporate sponsor such as a drug company may have directly supported it. If this is the case, the hospital needs to make this relationship clear so that the reader can understand that there may be a possible conflict of interest.

Commercial Businesses.   When a business offers information on its own Web site, the questions that need to be asked are somewhat different. Some information, such as software documentation and product pricing, will be objective. However, users must not assume that all the information will be objective, because the company's goal is to promote its own products and services. Therefore, readers need to ask the following questions:

- What is the company's purpose for offering the information?
- How are the products the company is promoting related to the information being provided?
- Are there offers for free products in exchange for some type of information from the user?
- If free entertainment is provided, what relation does it have to the products or services being offered?
- Is the business withholding more detailed information that is only available for a fee?
- Is marketing information being gathered, and for what purpose?

Strategies for Analyzing Web Information Provided by Sites That Have Advertisers or Sponsors

The following are three strategies that can be helpful when sorting out the relations between advertisers, sponsors, and information on the Web.

Identify the Key Players Involved in Providing Information at the Site.   Figure 3.6, the OncoLink Home Page, is an example of a site that clearly identifies some of the key players involved in providing the site's information. We can identify OncoLink as a subsite of the University of Pennsylvania Cancer Center, which in turn is a part of the University of Pennsylvania. The site clearly indicates its relationship to both these organizations and also provides information about its corporate and nonprofit sponsors. (Note: Figs. 3.7 and 3.8 utilize the method of annotating screen captures with important concepts that is used throughout the rest of the book.)

Figure 3.7, the home page of the radio show *Prairie Home Companion*, also identifies key players involved in providing the information at the site. The page clearly indicates that the parent organization is Minnesota Public Radio, and that the site has two corporate sponsors, Lands' End and Netscape. (The third sponsor, Guy's Shoes, is not quite so clearly indicated because Guy's Shoes is the fictitious sponsor of the radio show.)

WWW.CIOLEK.COM Asia Pacific Research Online is another site that identifies not only the key players involved in providing the information for the site, but also clearly states its policy concerning advertising (see Fig. 3.8). The policy indi-

University of Pennsylvania (the parent site)

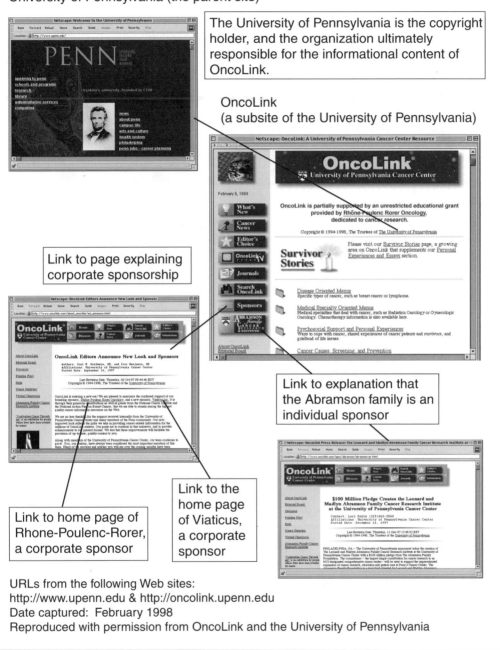

The University of Pennsylvania is the copyright holder, and the organization ultimately responsible for the informational content of OncoLink.

OncoLink
(a subsite of the University of Pennsylvania)

Link to page explaining corporate sponsorship

Link to explanation that the Abramson family is an individual sponsor

Link to home page of Rhone-Poulenc-Rorer, a corporate sponsor

Link to the home page of Viaticus, a corporate sponsor

URLs from the following Web sites:
http://www.upenn.edu & http://oncolink.upenn.edu
Date captured: February 1998
Reproduced with permission from OncoLink and the University of Pennsylvania

FIG. 3.6. Determining corporate and nonprofit sponsorship.

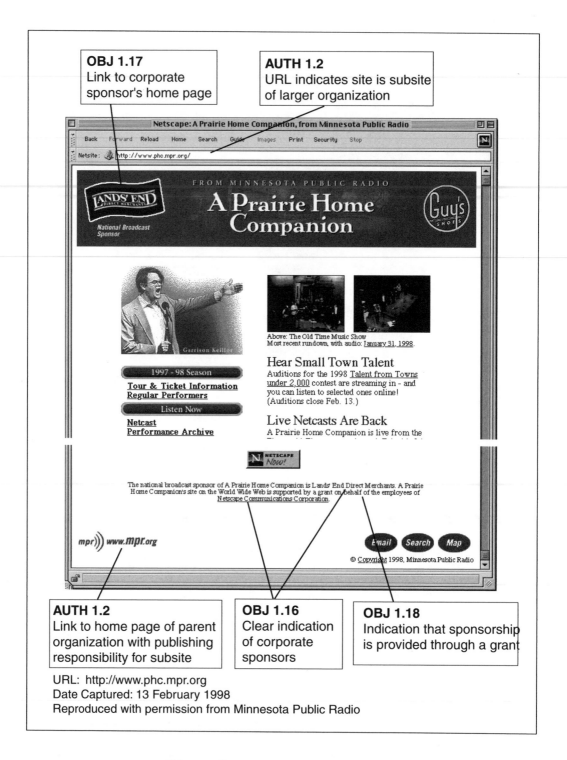

OBJ 1.17
Link to corporate sponsor's home page

AUTH 1.2
URL indicates site is subsite of larger organization

AUTH 1.2
Link to home page of parent organization with publishing responsibility for subsite

OBJ 1.16
Clear indication of corporate sponsors

OBJ 1.18
Indication that sponsorship is provided through a grant

URL: http://www.phc.mpr.org
Date Captured: 13 February 1998
Reproduced with permission from Minnesota Public Radio

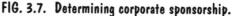

FIG. 3.7. Determining corporate sponsorship.

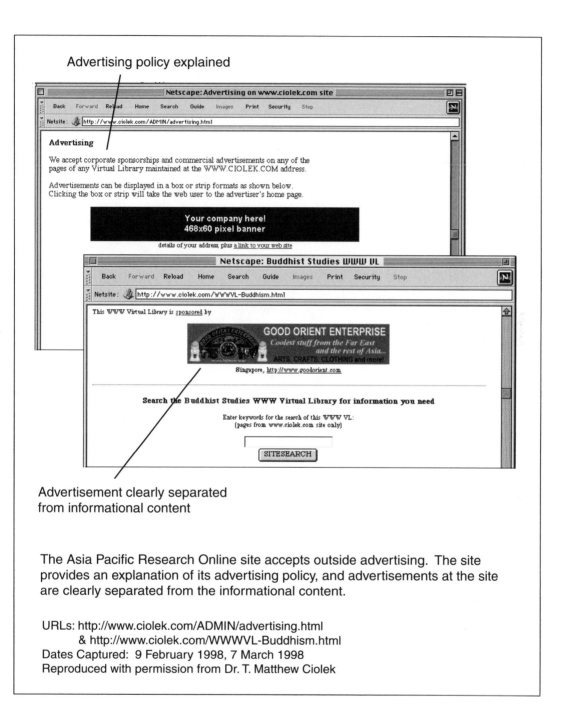

Advertising policy explained

Netscape: Advertising on www.ciolek.com site

Back Forward Reload Home Search Guide Images Print Security Stop

Netsite: http://www.ciolek.com/ADMIN/advertising.html

Advertising

We accept corporate sponsorships and commercial advertisements on any of the pages of any Virtual Library maintained at the WWW.CIOLEK.COM address.

Advertisements can be displayed in a box or strip formats as shown below. Clicking the box or strip will take the web user to the advertiser's home page.

Your company here!
468x60 pixel banner

details of your address, plus a link to your web site

Netscape: Buddhist Studies WWW VL

Back Forward Reload Home Search Guide Images Print Security Stop

Netsite: http://www.ciolek.com/WWWVL-Buddhism.html

This WWW Virtual Library is sponsored by

GOOD ORIENT ENTERPRISE
*Coolest stuff from the Far East
and the rest of Asia...*
ARTS, CRAFTS, CLOTHING and more!

Singapore, http://www.goodorient.com

Search the Buddhist Studies WWW Virtual Library for information you need

Enter keywords for the search of this WWW VL:
(pages from www.ciolek.com site only)

SITESEARCH

Advertisement clearly separated
from informational content

The Asia Pacific Research Online site accepts outside advertising. The site provides an explanation of its advertising policy, and advertisements at the site are clearly separated from the informational content.

URLs: http://www.ciolek.com/ADMIN/advertising.html
      & http://www.ciolek.com/WWWVL-Buddhism.html
Dates Captured: 9 February 1998, 7 March 1998
Reproduced with permission from Dr. T. Matthew Ciolek

FIG. 3.8. Example of commercial advertising clearly differentiated from informational content.

cates that any advertising or sponsorship at the site will be carefully differentiated from the informational content of the site.

Identify What Information at the Site Is in Actuality Advertising.

An analysis of what appears to be objective information at a business site may reveal that the information is biased in favor of the company. Figure 3.9 is an example of a site that provides a link to a supposedly objective bibliography about evaluating Web resources. However, a closer examination of the page reveals that one of the entries on this supposedly objective list of resources is in reality a link to a page designed to promote the company's services. By placing this link in the midst of links to legitimate evaluation sources, the company hopes to confer legitimacy on its own site.

Identify the Purpose for Providing Entertainment at the Site.

Some business sites provide entertainment as a way of drawing in users so they can be given a marketing message. Figure 3.9 provides an example of a site that blends entertainment and advertising in this way.

CONCLUSION

This chapter has presented concepts it is important to understand when trying to sort out the relations among advertisers, sponsors, and Web information. The chapter has also stressed the importance of understanding the influence these relations might have on the objectivity of the information. However, just because a site includes advertising does not necessarily mean that the information contained at the site is not objective. Similarly, an absence of advertising does not guarantee that the material at the site is without bias. As Dr. Ciolek (personal communication, March 27, 1998) pointed out, the objectivity and trustworthiness of information at a site are not so much the result of the financial backing for the site, but rather a function of the site owner's professionalism and integrity. Therefore, in assessing the trustworthiness of a site, it is not sufficient to just determine who the advertisers and sponsors are. It is also important to assess the trustworthiness and authority of the person, organization, or business responsible for the information at the site. The next seven chapters include numerous tools and techniques to aid a Web user in analyzing the potential trustworthiness of information found at a Web site.

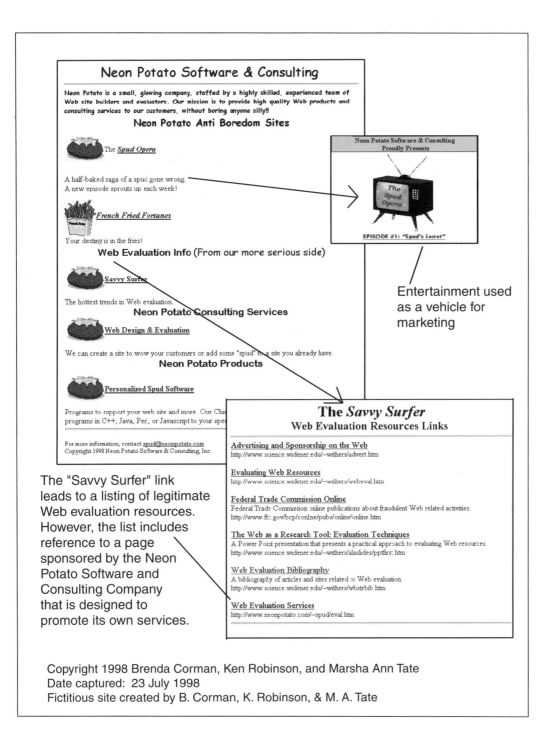

Neon Potato Software & Consulting

Neon Potato is a small, glowing company, staffed by a highly skilled, experienced team of Web site builders and evaluators. Our mission is to provide high quality Web products and consulting services to our customers, without boring anyone silly!!

Neon Potato Anti Boredom Sites

The *Spud Opera*

A half-baked saga of a spud gone wrong.
A new episode sprouts up each week!

French Fried Fortunes

Your destiny is in the fries!

Web Evaluation Info (From our more serious side)

Savvy Surfer

The hottest trends in Web evaluation.

Neon Potato Consulting Services

Web Design & Evaluation

We can create a site to wow your customers or add some "spud" to a site you already have.

Neon Potato Products

Personalized Spud Software

Programs to support your web site and more. Our Chie[f]
programs in C++, Java, Per[l], or Javascript to your spe[c]

For more information, contact spud@neonpotato.com
Copyright 1998 Neon Potato Software & Consulting, Inc.

Neon Potato Software & Consulting
Proudly Presents

The Spud Opera

EPISODE #1: *"Spud's Secret"*

Entertainment used as a vehicle for marketing

The *Savvy Surfer*
Web Evaluation Resources Links

Advertising and Sponsorship on the Web
http://www.science.widener.edu/~withers/advert.htm

Evaluating Web Resources
http://www.science.widener.edu/~withers/webeval.htm

Federal Trade Commission Online
Federal Trade Commission online publications about fraudulent Web related activities.
http://www.ftc.gov/bcp/conline/pubs/online/online.htm

The Web as a Research Tool: Evaluation Techniques
A Power Point presentation that presents a practical approach to evaluating Web resources.
http://www.science.widener.edu/~withers/alaslides/pptfirst.htm

Web Evaluation Bibliography
A bibliography of articles and sites related to Web evaluation.
http://www.science.widener.edu/~withers/wbstrbib.htm

Web Evaluation Services
http://www.neonpotato.com/~spud/eval.htm

The "Savvy Surfer" link leads to a listing of legitimate Web evaluation resources. However, the list includes reference to a page sponsored by the Neon Potato Software and Consulting Company that is designed to promote its own services.

Copyright 1998 Brenda Corman, Ken Robinson, and Marsha Ann Tate
Date captured: 23 July 1998
Fictitious site created by B. Corman, K. Robinson, & M. A. Tate

FIG. 3.9. Example of a site that blends information, advertising, and entertainment.

4

Applying Basic Evaluation
Criteria to a Web Page

Chapter Contents

- **How to Use Chapters 4 Through 10**
- **Incorporation of the Basic Elements Into Web Pages**
- **The Importance of Interaction and Transaction Features**
- **An Introduction to Navigational Aids and Nontext Features**
- **Information on the Six Types of Web Pages**
- **The Checklist of Basic Elements: Keys to Evaluating or Creating Web Pages**

HOW TO USE CHAPTERS 4 THROUGH 10

This chapter introduces basic elements important to include on any Web page. The chapter also identifies six different types of pages (advocacy, business, informational, news, personal, and entertainment) and discusses the need to include additional elements on each of these types of pages.

When using chapters 4 through 10 to either evaluate or create Web pages:

1. Read chapter 4 to learn the basic elements that need to be included on any page, regardless of its type. Also, review the Checklist of Basic Elements located at the end of the chapter.
2. Determine what type of page you are evaluating or creating. Each of the following six chapters is devoted to one of the six types of pages and begins with information to assist the user in identifying that type of page.
3. Consult the appropriate chapter to learn what elements, in addition to the basic ones listed in this chapter, need to be included when evaluating or creating this unique type of page.

36

INCORPORATION OF THE BASIC ELEMENTS INTO WEB PAGES

This section discusses key Web page elements and provides illustrations of how they have been incorporated into actual Web pages. The topics covered are:

- Authority (site level).
- Authority (page level).
- Accuracy.
- Objectivity.
- Currency.
- Coverage and intended audience

Authority (Elements 1 and 2)

Authority is the extent to which material is the creation of a person or organization recognized as having definitive knowledge of a given subject area. When discussing the authority of information on the Web, it is first helpful to analyze the authority of the Web site as a whole, and then the authority of an individual page within the site.

Element 1: Authority (Site Level)

One of the most important aspects of evaluating a Web site as a whole is ascertaining the authority of the site. Figure 4.1 is an illustration of the home page from the OncoLink site at the University of Pennsylvania. From an analysis of this home page, we can determine the following factors:

- The OncoLink and University of Pennsylvania Cancer Center symbols appear together prominently at the top of the page. By following the link to "About OncoLink" we discover that the University of Pennsylvania Cancer Center is the organization directly responsible for the information at the OncoLink site.
- The page is copyrighted by the Trustees of the University of Pennsylvania, the organization that appears to have ultimate publishing responsibility for the site.
- A link is provided to the University of Pennsylvania home page, which is the parent organization for OncoLink.
- An Editorial Board oversees the contents of the site.
- An e-mail address is provided as a way of contacting OncoLink.

Authority, Site Level: Questions to Ask. The following questions are important to consider when establishing the authority of a Web site. The items referred to in the following questions should be located either on the page itself or on a page directly linked to it.

- Is it clear what organization, company, or person is responsible for the contents of the site? This can be indicated via a logo. Without this basic information, it is virtually impossible to verify the authority of the site.

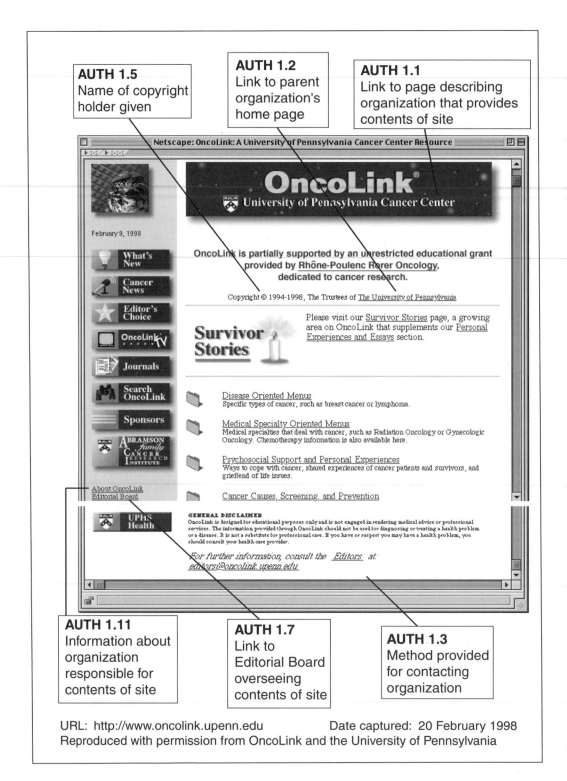

AUTH 1.5
Name of copyright holder given

AUTH 1.2
Link to parent organization's home page

AUTH 1.1
Link to page describing organization that provides contents of site

AUTH 1.11
Information about organization responsible for contents of site

AUTH 1.7
Link to Editorial Board overseeing contents of site

AUTH 1.3
Method provided for contacting organization

URL: http://www.oncolink.upenn.edu   Date captured: 20 February 1998
Reproduced with permission from OncoLink and the University of Pennsylvania

FIG. 4.1. Keys to verifying authority (site level).

- If the site is a subsite of a larger organization, does the site provide the logo or name of the larger organization? Knowing the name of a parent organization that has ultimate responsibility for the contents of a subsite can help add legitimacy to the material at the subsite.
- Is there a way to contact the organization, company, or person responsible for the contents of the site? These contact points can be used to verify the legitimacy of the site. Although a phone number, a mailing address, and an e-mail address are all possible contact points, a mailing address is critical if there is any question about the legitimacy of the site.
- Are the qualifications of the organization, company, or person responsible for the contents of the site indicated? Including such qualifications is especially important if the site does not originate from a well-known source.
- If all the materials on the site are protected by a single copyright holder, is the name of the copyright holder given? The copyright holder is often the same as the contact point, but if not, as in the OncoLink example in Fig. 4.1, it can provide additional information about the authority of the site.
- Does the site list any recommendations or ratings from outside sources?

Element 2: Authority (Page Level)

To evaluate the authority of a Web page, you must look at both the authority of the page itself and the authority of the site on which the page resides. Before looking at the authority of the page, it is helpful to first return to the site's home page to analyze the authority of the site.

Figure 4.2, "Nurse's Notes: Second Hand Smoke," is an example of a page that provides considerable information about the authority of the person and organization offering the information. For users unfamiliar with OncoLink, the URL address, http://www.upenn.oncolink.edu, would initially provide few clues about the authority of the page unless the user were already familiar with the University of Pennsylvania's URL. However, additional information is readily available from a number of other sources on the page.

The OncoLink and University of Pennsylvania Cancer Center symbols appear together prominently at the top of the page. They also serve as links back to the OncoLink home page that, if followed, reveal that the University of Pennsylvania Cancer Center is the organization directly responsible for the contents of OncoLink. In addition, the page's information is copyrighted by The Trustees of the University of Pennsylvania, and an e-mail address is also provided for OncoLink. Moreover, the OncoLink pages follow a consistent design throughout the site. This design consistency not only helps users recognize that they are still within the OncoLink site, but also serves as an additional aid to verifying the authority of the page.

The informational contents of the page are attributed to Maggie Hampshire, whose academic qualifications and affiliation are clearly indicated. Using the OncoLink internal search engine, a query for "Maggie Hampshire" results in a list of other articles she has written, as well as a link to her biography. Another method for assisting in the verification of an author's qualifications is to perform a search on one of the general Web search engines such as AltaVista or Infoseek to see what additional information about the author can be located.

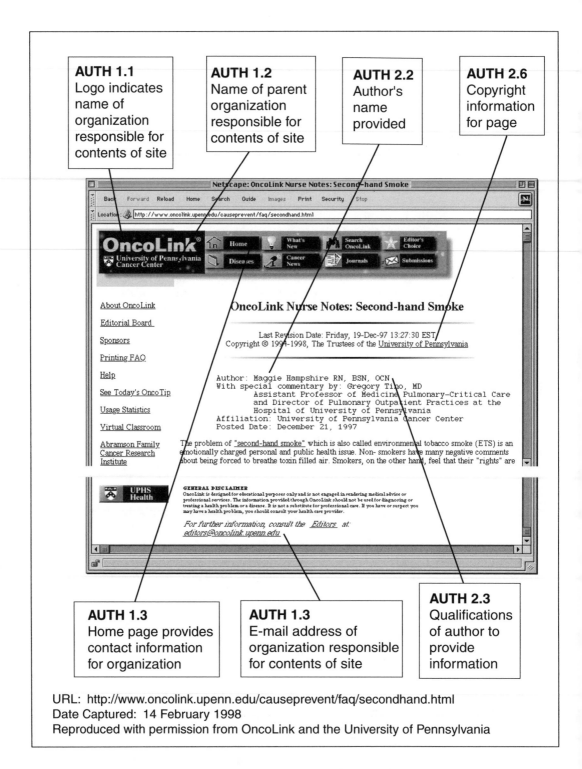

AUTH 1.1
Logo indicates name of organization responsible for contents of site

AUTH 1.2
Name of parent organization responsible for contents of site

AUTH 2.2
Author's name provided

AUTH 2.6
Copyright information for page

AUTH 1.3
Home page provides contact information for organization

AUTH 1.3
E-mail address of organization responsible for contents of site

AUTH 2.3
Qualifications of author to provide information

URL: http://www.oncolink.upenn.edu/causeprevent/faq/secondhand.html
Date Captured: 14 February 1998
Reproduced with permission from OncoLink and the University of Pennsylvania

FIG. 4.2. Keys to verifying authority (page level).

Authority, Page Level: Questions to Ask. The following questions are important to consider when ascertaining the authority of an individual Web page. The items referred to in these questions should be located either on the page itself or on a page directly linked to it.

- Is it clear what organization, company, or person is responsible for the contents of the page? For a page written by an individual with no organizational affiliation, it is important to indicate responsibility for the page.
- If the material on the page is written by an individual, are the author's name and qualifications for providing the information clearly stated? Even though the author of the page in Fig. 4.2 is affiliated with an institution, and the page has the official approval of that institution, listing her qualifications for writing the material gives the page added authority.
- Is there a way of contacting the author? That is, does the person list a phone number, mailing address, or e-mail address? These contact points can be an important way of verifying that an individual is who they say they are.
- Is there a way of verifying the author's qualifications? That is, is there an indication of his or her expertise in the subject area, or a listing of memberships in professional organizations related to the topic?
- If the material on the page is copyright protected, is the name of the copyright holder given? As with the copyright holder for a site, the copyright holder for the page is another indication of who has ultimate responsibility for the contents of the page.
- Does the page have the official approval of the company, organization, or person responsible for the site? For pages at the site of a business or organization, as in the OncoLink example in Fig. 4.2, a similarity in page layout and design features often indicates that the page has the official approval of the site. If the page does not have such official approval, there will often be a disclaimer stating so, as is frequently the case with student pages at a college or university Web site.

The OncoLink example in Fig. 4.2 provides a clear indication of the authority of the page. There may be times, however, when a Web page has been retrieved by a search engine and there is no indication on the page of who has responsibility for it. In such a situation, the following strategies may assist a user in determining the source of the information:

- If possible, return to the home page, because it is usually one of the best sources for discovering what person or organization is responsible for the contents of the page.
- Analyze the URL address of the page to see if it offers clues as to who is responsible for the information on the page.
- Attempt to truncate the URL address by removing the end of it to determine if the page that contains the original link to the page being evaluated can be retrieved. For example, if the page's URL is http://www.host.com/~jsmith/cc17.htm, delete the file name *cc17.htm* and attempt to go the linking page.

Element 3: Accuracy of the Information

Accuracy is the extent to which information is reliable and free from errors.

The page "Nurse's Notes: Second Hand Smoke" (Fig. 4.3) includes several useful indicators to help determine the accuracy of the information provided. First, the page is free of spelling and typographical errors. This fact does not assure the accuracy of the contents, but pages free of errors in spelling, punctuation, and grammar do indicate that care has been taken in producing the page. Most important, Ms. Hampshire has enabled readers to verify the factual information contained in the article. Not only are the sources of the factual information named, but links are also provided to many of the original sources (many of which are at government sites). In addition, it is clear that an Editorial Board has ultimate responsibility for the accuracy of the information provided.

Accuracy: Questions to Ask.   The following are important questions to consider when determining the accuracy of a Web page.

- Is the information free of grammatical, spelling, and typographical errors? As already stated, these types of errors not only indicate a lack of quality control, but can actually produce inaccuracies in information.
- Are sources for factual information provided so the facts can be verified in the original source? A user needs to both verify that authoritative sources have been used to research the material and also be able to access the sources cited if desired.
- If there are any graphs, charts, or tables, are they clearly labeled and easy to read? Legibility is a critical element to consider when converting graphs, charts, or tables into electronic form.

Element 4: Objectivity of the Information

Objectivity is the extent to which material expresses facts or information without distortion by personal feelings or other biases.

The objectivity of the information at the OncoLink site can be evaluated in several ways. Because of the possibility of influence by an advertiser or sponsor on the objectivity of the information, it is important to first look at any advertising or sponsorship present. An analysis of the OncoLink home page (Fig. 4.4) clearly indicates the site has two corporate sponsors and provides a link to the home page of each of them. With this information, a user can learn more about the nature of the corporations that are donating money to the site. It is also useful to look at what type of sponsorship is being provided. The type of corporate sponsorship at the OncoLink site is clearly listed as an "unrestricted educational grant." The site also indicates it has a nonprofit sponsor and provides a link to an explanation of the nature of this nonprofit sponsorship.

In addition to examining the potential bias from an outside sponsor, it is also important to analyze to what degree the information provider intends to be objective. Although this can often be difficult to determine, particularly for individual

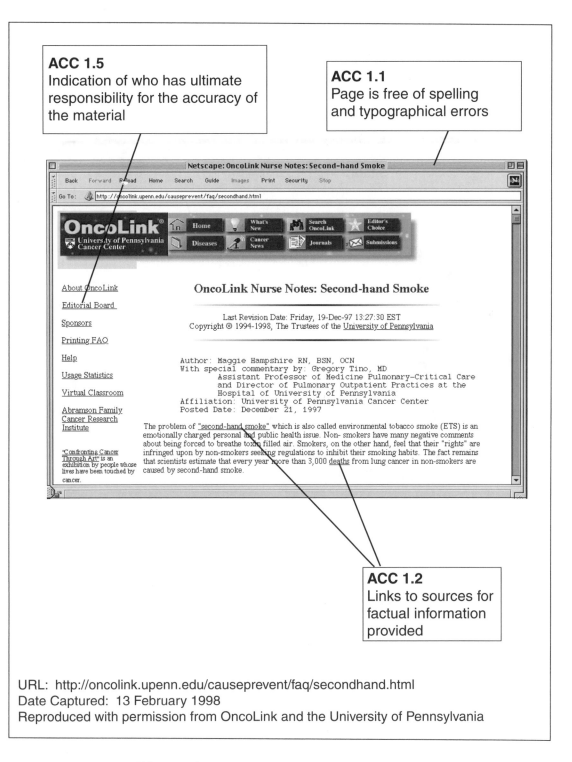

FIG. 4.3. Keys to verifying the accuracy of a Web page.

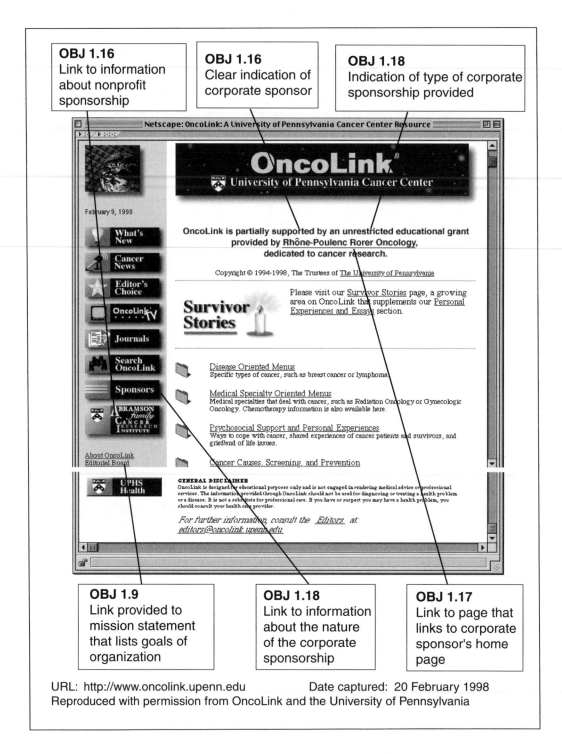

OBJ 1.16
Link to information about nonprofit sponsorship

OBJ 1.16
Clear indication of corporate sponsor

OBJ 1.18
Indication of type of corporate sponsorship provided

OBJ 1.9
Link provided to mission statement that lists goals of organization

OBJ 1.18
Link to information about the nature of the corporate sponsorship

OBJ 1.17
Link to page that links to corporate sponsor's home page

URL: http://www.oncolink.upenn.edu   Date captured: 20 February 1998
Reproduced with permission from OncoLink and the University of Pennsylvania

FIG. 4.4. Keys to verifying the objectivity of a Web site.

authors, when analyzing an organization's pages clues can sometimes be obtained by looking at the mission statement.

The "About OncoLink" link on the home page retrieves for users the "Mission Statement for OncoLink." The mission statement explains the rationale behind OncoLink and also conveys its stated objectives: "1. Dissemination of information relevant to the field of oncology. 2. Education of health care personnel. 3. Education of patients, families, and other interested parties, and 4. Rapid collection of information pertinent to the specialty." It is clear that the site is intended to present objective information for educational purposes.

Objectivity: Questions to Ask.   The following are questions important to consider when analyzing the objectivity of Web information.

- Is the point of view of the individual or organization responsible for providing the information evident? It is important to know to what degree the information provider is attempting to be objective with the information offered on the Web page.
- If there is an individual author for the material on the page, is the point of view of the author evident?
- If there is an author for the page, is it clear what relationship exists between the author and the person, company, or organization responsible for the site? It is important to know to what extent the organization responsible for the contents of the site might influence the objectivity of the author.
- Is the page free of advertising? If not, it is important to try to determine to what extent an advertiser might influence the informational contents.
- If there is advertising on the page, is it clear what relationship exists between the company, organization, or person responsible for the informational contents of the page and any advertisers represented on the page?
- If there is both advertising and information on the page, is there a clear differentiation between the two?
- Is there an explanation of the site's policy relating to advertising and sponsorship?
- If the site has nonprofit or corporate sponsors, are they clearly listed? If so, it is important to try to determine to what extent the sponsor might influence the informational content.
- Are links included to the sites of any nonprofit or corporate sponsors so a user can learn more about them?
- Is additional information provided about the nature of the sponsorship, such as an indication of what type it is (nonrestrictive, educational, etc.)?

Element 5: Currency of the Information

Currency is the extent to which material can be identified as up to date.

By analyzing the "Nurse Notes: Second Hand Smoke" page from OncoLink (Fig. 4.5), we can readily determine the currency of the page. It is clear that the page's contents were first posted to OncoLink on December 21, 1997, and a date and time of last revision are also included on the page. In addition, the date of last

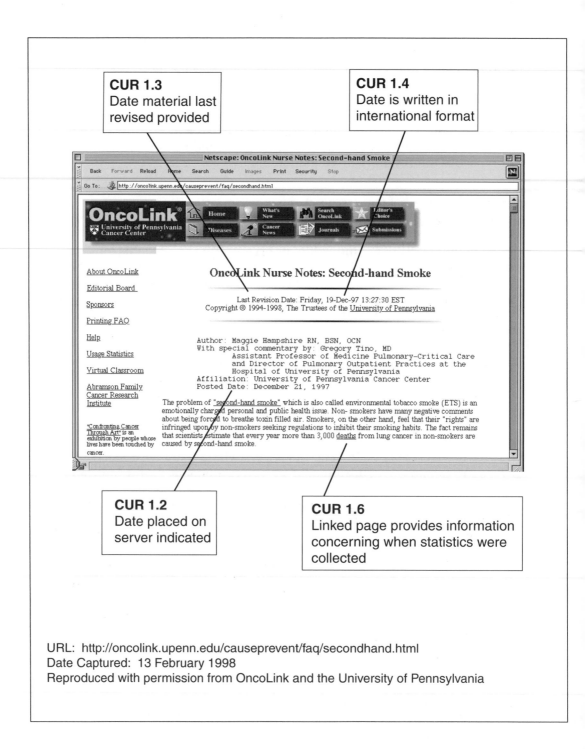

CUR 1.3
Date material last revised provided

CUR 1.4
Date is written in international format

CUR 1.2
Date placed on server indicated

CUR 1.6
Linked page provides information concerning when statistics were collected

URL: http://oncolink.upenn.edu/causeprevent/faq/secondhand.html
Date Captured: 13 February 1998
Reproduced with permission from OncoLink and the University of Pennsylvania

FIG. 4.5. Keys to verifying the currency of a Web page.

revision is in a format readily understood by international readers. For statistical information on the page, a link is provided to the original source of the information. The original source clearly indicates the date the statistics were collected.

Currency: Questions to Ask. The following is a list of questions important to consider when determining the currency of a Web page.

- Is the date the material was first created in any format included on the page?
- Is the date the material was first placed on the server included on the page?
- If the contents of the page have been revised, is the date (and time, if appropriate) the material was last revised included on the page?
- To avoid confusion, are all creation and revision dates in an internationally recognized format? Examples of dates in international format (dd mm yy) are 5 June 1997 and 21 January 1999.

Element 6: Coverage of the Information and Its Intended Audience

Coverage is the range of topics included in a work and the depth to which those topics are addressed. The *intended audience* is the group of people for whom the material was created.

The coverage of the OncoLink site can be ascertained in several ways. First, the "General Disclaimer" that appears at the bottom of the home page (Fig. 4.4) offers some insight into the topics covered. The disclaimer states "OncoLink is designed for educational purposes only and is not engaged in rendering medical advice or professional services."

In addition, the topics included at the site are listed on the home page, so the user can readily determine the types of information that will be found at the site. The mission statement for OncoLink (Fig. 4.6) also provides additional insight about the coverage of the site by listing the types of materials at the site and its stated objectives. The mission statement also indicates the intended audiences for the materials are health care professionals, patients, families, and other interested parties.

Coverage and Intended Audience: Questions to Ask. The following are questions to consider when determining the coverage and intended audience of a Web site.

- Is it clear what materials are included at the site? This can be difficult to determine unless there is an index to the site or a site map.
- If the page is still under construction, is the expected date of completion indicated?
- Is the intended audience for the material clear?
- If material is presented for several different audiences, is the intended audience for each type of material clear?

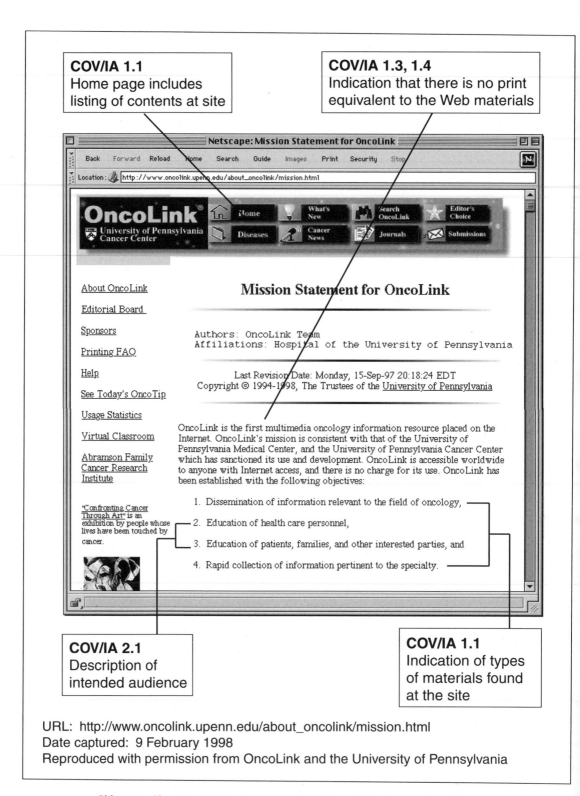

COV/IA 1.1
Home page includes listing of contents at site

COV/IA 1.3, 1.4
Indication that there is no print equivalent to the Web materials

Netscape: Mission Statement for OncoLink

Back Forward Reload Home Search Guide Images Print Security Stop

Location: http://www.oncolink.upenn.edu/about_oncolink/mission.html

OncoLink®
University of Pennsylvania
Cancer Center

Home | What's New | Search OncoLink | Editor's Choice
Diseases | Cancer News | Journals | Submissions

About OncoLink

Editorial Board

Sponsors

Printing FAQ

Help

See Today's OncoTip

Usage Statistics

Virtual Classroom

Abramson Family Cancer Research Institute

"Confronting Cancer Through Art" is an exhibition by people whose lives have been touched by cancer.

Mission Statement for OncoLink

Authors: OncoLink Team
Affiliations: Hospital of the University of Pennsylvania

Last Revision Date: Monday, 15-Sep-97 20:18:24 EDT
Copyright © 1994-1998, The Trustees of the University of Pennsylvania

OncoLink is the first multimedia oncology information resource placed on the Internet. OncoLink's mission is consistent with that of the University of Pennsylvania Medical Center, and the University of Pennsylvania Cancer Center which has sanctioned its use and development. OncoLink is accessible worldwide to anyone with Internet access, and there is no charge for its use. OncoLink has been established with the following objectives:

1. Dissemination of information relevant to the field of oncology,

2. Education of health care personnel,

3. Education of patients, families, and other interested parties, and

4. Rapid collection of information pertinent to the specialty.

COV/IA 2.1
Description of intended audience

COV/IA 1.1
Indication of types of materials found at the site

URL: http://www.oncolink.upenn.edu/about_oncolink/mission.html
Date captured: 9 February 1998
Reproduced with permission from OncoLink and the University of Pennsylvania

FIG. 4.6. Keys to verifying the coverage and intended audience of a Web site.

THE IMPORTANCE OF INTERACTION AND TRANSACTION FEATURES

Interaction and transaction features are an additional category of basic elements important to include on any type of Web page, no matter what its type. Interaction and transaction features are tools that enable a user to interact with the person or organization responsible for a Web site, or enter into a transaction (usually financial) via a Web site.

Some ways a user might interact with a site are obvious—for example, filling out an online order form, or providing information such as a credit card number. Other ways a user might provide information to the person or organization responsible for the site are not so obvious (and may in fact be transparent to the user), as when a site collects information about a user via cookies.

Cookies, as described by Rankin (1998), enable data to be stored by a server on a user's computer and therefore allow information about a user's visit to a particular site to be saved. The information stored in a cookie cannot be read back later by any site except the one that originally supplied the cookies. Some of the features made possible by the use of cookies, as described by Rankin, include:

- The storage of information such as a user ID and password, thereby eliminating the need for the information to be entered again when the user returns to a site.
- The creation of a "shopping cart" into which a user can place items before paying for them.
- The tracking by advertisers of the pages a user visits. This enables advertisers to tailor ads to the user's interests and to monitor the effectiveness of the pages.
- The personalization of a Web site according to a user's preferences.

Although cookies may expire when a user leaves the browser, typically they are stored on the user's computer, and they can be read by the server when the user visits the site again.

Whether a site is collecting information openly by such means as order forms, or transparently using mechanisms such as cookies, it is important that users have confidence that the information they are providing to the site will be kept confidential unless the customer indicates it may be made public. Therefore, it is important that the site make clear its policy regarding the confidentiality of information collected, both while it is in transit to the site, and also once it has arrived at the site. This can be done not only by stating the site's policy on these issues, but also by indicating what technical measures the site has in place to ensure such privacy.

Other important interaction and transaction features concern how easily the user can provide feedback to the site, and restrictions on the use of the materials offered at the site.

The following is a list of important general considerations concerning interaction and transaction features:

- If any financial transactions occur at the site, does the site indicate what means have been taken to ensure their security? A secure transaction is an encrypted, or scrambled, communication between a Web server and a browser. Because the data communicated in a secure transaction are encrypted, the opportunity for the content to be read by an unauthorized person during the transfer across the Web is minimized.

- If the business, organization, or person responsible for the page is requesting information from the user, is there a clear indication of how it will be used?
- If cookies are used at the site, is the user notified? Is there an indication of what the cookies are used for and how long they last?
- Is there a feedback mechanism for users to comment about the site?
- Are any restrictions regarding downloading and other uses of the materials offered on the page clearly stated?

AN INTRODUCTION TO NAVIGATIONAL AIDS AND NONTEXT FEATURES

In addition to the basic elements previously described, there are two additional features that need to be considered in the evaluation and creation of Web pages: navigational aids and nontext features. Checklists for these two categories of features are included in chapter 11, which is devoted exclusively to issues involved in creating effective Web pages and sites. However, this chapter defines these elements and provides examples of how they are used because, although they are critical factors in the creation of Web sites, they also play an important, if more indirect, role in the evaluation of Web information.

Consistent and Effective Use of Navigational Aids

Navigational aids are elements that help a user locate information at a Web site and allow the user to easily move from page to page within the site. Navigational aids allow readers the flexibility they need to move to a desired spot in the Web site. They are necessary for two reasons: (a) they allow readers to "browse" easily through the site, and (b) they provide an orientation to the material at the site, just as page numbers, chapter headings, a table of contents, and an index provide an orientation to material within a book.

The following navigational aids are important to include on Web pages and/or sites:

- A browser title.
- A page title.
- Internal links within a site.
- The URL for the page.
- A site map or index.
- An internal search engine for the site (if appropriate).

Browser Title

The *browser title* is the title of a page that is picked up by the browser from the HTML <TITLE> Tag. It has the following characteristics:

- It usually appears as part of the browser frame at the top of the browser window (in the "title bar").
- It is distinguished from the page title, which is the title that appears in the body of the Web page.
- The presence of the browser title allows the user to quickly identify the contents of the page.

Browser Title Examples   The following is an example of a home page browser title that clearly indicates both the company responsible for the page and also that the page is the home page:

> *The Neon Potato Company Home Page*

The following are examples of non-home page browser titles that clearly indicate both the company responsible for the page and the specific contents of the page:

> *Neon Potato: Company Information*

> *Neon Potato: Product Information*

> *Neon Potato: Copyright Information*

Additional Points About Browser Titles.   Whatever appears in the HTML <TITLE> Tag will typically become the default title of any browser bookmark to the page. The browser title should be descriptive of the page's contents so it can be easily recognized in a bookmark list. (A bookmark is a URL address stored on a user's computer that allows the user to easily return to a frequently used page. The ability to store bookmarks is a common browser capability.)

Whatever appears in the HTML <TITLE> Tag is usually picked up by a search engine and used as the default description of that page. Therefore, each browser title should be concise, yet descriptive. When creating browser titles, it may be helpful to think of the title of a site as the title of a book, and the title of a page as a chapter heading in a book.

Page Title

The *page title* is the title found in the text of the Web page. The presence of a page title allows the user to quickly identify the contents of the page. It is often created by using an HTML Heading (usually an <H1>) Tag. The page title will frequently be the same as the browser title for the page.

URL for the Page

The *URL* (Uniform Resource Locator) is an identifier that uniquely distinguishes the page from all other World Wide Web pages. Including the URL in the body of the page enables users who print out the page to have a printed record of its source and revisit the page at a later date.

Hypertext Links

Hypertext links (or simply *links*) are regions of a Web page that, when selected, cause a different Web page or a different part of the same Web page to be displayed. A link can consist of a word or phrase of text, or an image. The inclusion of links on a Web page allows users to move easily from one Web page to another.

Including appropriate links allows the user to navigate within the site according to individual preferences, and to return to the site's home page, site map, or (for sites arranged in a hierarchy) to the page one level up in the hierarchy. Having the ability to return to the site's home page is important in the process of determining the authority of the site, and having the ability to access the site map is important in determining the coverage of the site. In addition, it is helpful to have internal links placed in a consistent place on each page and be uniform in appearance, whether they are created with graphics or text.

Site Map and Index

A *site map* is a display, often graphical, of the major components of a Web site. An *index* is a listing, often alphabetical, of the major components of a Web site.

A site map or index provides a quick overview of the pages contained within the entire site, and each can be an important tool in determining the coverage of the site.

Internal Search Engine

In contrast to the well-known Web search engines such as AltaVista, Infoseek, and HotBot that search for words or phrases on a large number of Web pages, an *internal search engine* is one that searches for words or phrases only within one World Wide Web site.

An internal search engine is a helpful navigational aid for sites that present large amounts of information, as it allows users to locate information at the site quickly and easily.

Figure 4.7 illustrates how navigational aids have been incorporated into the OncoLink Web site to improve its functionality.

Effective Use of Nontext Features

Nontext features include those elements that require the user to have additional software or a specific browser to utilize the contents of a Web page. Some examples of nontext features include graphics, image maps, sound, and video.

When nontext features are present at a site, some users may not have the ability to take advantage of them. Users may, for example, be viewing the site with:

- A text-only browser.
- The browser's ability to display graphics turned off.
- Special software designed for those who are visually or physically disabled.

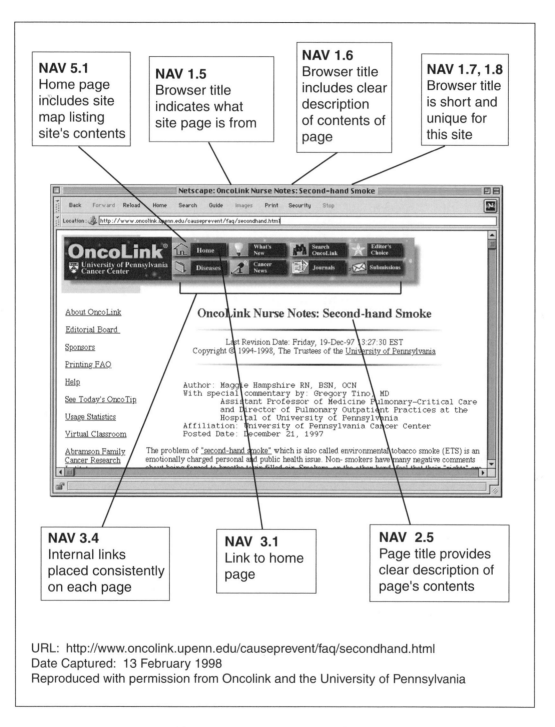

NAV 5.1
Home page includes site map listing site's contents

NAV 1.5
Browser title indicates what site page is from

NAV 1.6
Browser title includes clear description of contents of page

NAV 1.7, 1.8
Browser title is short and unique for this site

NAV 3.4
Internal links placed consistently on each page

NAV 3.1
Link to home page

NAV 2.5
Page title provides clear description of page's contents

URL: http://www.oncolink.upenn.edu/causeprevent/faq/secondhand.html
Date Captured: 13 February 1998
Reproduced with permission from Oncolink and the University of Pennsylvania

FIG. 4.7. Examples of navigational aids.

Therefore, some important considerations for the use of nontext features include the following:

- If the page includes a graphic such as a logo or an image map, is there a text alternative for those viewing the page in text-only mode?
- If the page includes a nontext file (such as a sound or video file) that may require additional software to utilize it, is there an indication of the additional software needed and where it can be obtained?
- If a file requires additional software to access it, wherever possible is the same information provided in another format that does not require the additional software?
- If a page requires a specific browser or a specific version of a browser, does the page specify what is needed and indicate where it can be obtained?
- When following a link results in the loading of a large graphic, sound, or video file, is information provided to alert the user that this will happen?

INFORMATION ON THE SIX TYPES OF WEB PAGES

The first seven elements discussed in this chapter have been compiled into a Checklist of Basic Elements that can be used as a basis for evaluating or creating any Web page, no matter what its type. For a discussion of additional issues that need to be addressed when creating or evaluating advocacy, business, informational, news, personal, and entertainment pages, consult chapters 5 through 10.

For the reader's convenience all of the book's checklists have been assembled together in Appendix A.

Whether you are evaluating existing Web pages or creating new ones, it is important to analyze them page by page rather than assuming that all pages at a given site will be of one type. For example, it is common to find both informational and advocacy pages at the same site, and also common to find sites that have business pages combined with entertainment pages. Also, personal Web sites often combine different types of pages. It is not uncommon for a personal site to include entertainment items about a favorite musician, provide information about a favorite research topic, advocate a favorite cause, and try to sell a used bicycle all at the same time.

Not only can a site contain pages of different types, but individual pages can also be a combination of several different types. Such combination pages may require the use of additional checklists as appropriate.

THE CHECKLIST OF BASIC ELEMENTS: KEYS TO EVALUATING OR CREATING WEB PAGES

The following questions are general ones that need to be asked when evaluating or creating any Web page, no matter what its type. Answering these questions will help a user determine whether the information on a Web page is coming from an authoritative, accurate, and reliable source. The greater the number of "yes" answers, the greater the likelihood that the quality of the information on the page can be determined. The questions can also be used by Web authors as a guide to creating pages that can be recognized as originating from a reliable, trustworthy source.

Authority (AUTH)

Authority of a Site

The following information should be included either on a site's home page or on a page directly linked to it.

- ❏ Is it clear what organization, company, or person is responsible for the contents of the site? This can be indicated by the use of a logo. **AUTH 1.1**
- ❏ If the site is a subsite of a larger organization, does the site provide the logo or name of the larger organization? **AUTH 1.2**
- ❏ Is there a way to contact the organization, company, or person responsible for the contents of the site? These contact points can be used to verify the legitimacy of the site. Although a phone number, mailing address, and e-mail address are all possible contact points, a mailing address and phone number provide a more reliable way of verifying legitimacy. **AUTH 1.3**
- ❏ Are the qualifications of the organization, company, or person responsible for the contents of the site indicated? **AUTH 1.4**
- ❏ If all the materials on the site are protected by a single copyright holder, is the name of the copyright holder given? **AUTH 1.5**
- ❏ Does the site list any recommendations or ratings from outside sources? **AUTH 1.6**

Authority of a Page

- ❏ Is it clear what organization, company, or person is responsible for the contents of the page? Similarity in page layout and design features can help signify responsibility. **AUTH 2.1**

If the material on the page is written by an individual author:

- ❏ Is the author's name clearly indicated? **AUTH 2.2**
- ❏ Are the author's qualifications for providing the information stated? **AUTH 2.3**
- ❏ Is there a way of contacting the author? That is, does the person list a phone number, mailing address, and e-mail address? **AUTH 2.4**
- ❏ Is there a way of verifying the author's qualifications? That is, is there an indication of his or her expertise in the subject area, or a listing of memberships in professional organizations related to the topic? **AUTH 2.5**
- ❏ If the material on the page is copyright protected, is the name of the copyright holder given? **AUTH 2.6**
- ❏ Does the page have the official approval of the person, organization, or company responsible for the site? **AUTH 2.7**

Accuracy (ACC)

❑ Is the information free of grammatical, spelling, and typographical errors? **ACC 1.1**
❑ Are sources for factual information provided, so the facts can be verified in the original source? **ACC 1.2**
❑ If there are any graphs, charts, or tables, are they clearly labeled and easy to read? **ACC 1.4**

Objectivity (OBJ)

❑ Is the point of view of the individual or organization responsible for providing the information evident? **OBJ 1.1**

If there is an individual author of the material on the page:

❑ Is the point of view of the author evident? **OBJ 1.2**
❑ Is it clear what relationship exists between the author and the person, company, or organization responsible for the site? **OBJ 1.3**
❑ Is the page free of advertising? **OBJ 1.4**

For pages that include advertising:

❑ Is it clear what relationship exists between the business, organization, or person responsible for the contents of the page and any advertisers represented on the page? **OBJ 1.5**
❑ If there is both advertising and information on the page, is there a clear differentiation between the two? **OBJ 1.6**
❑ Is there an explanation of the site's policy relating to advertising and sponsorship? **OBJ 1.7**

For pages that have a nonprofit or corporate sponsor:

❑ Are the names of any nonprofit or corporate sponsors clearly listed? **OBJ 1.16**
❑ Are links included to the sites of any nonprofit or corporate sponsors so that a user may find out more information about them? **OBJ 1.17**
❑ Is additional information provided about the nature of the sponsorship, such as what type it is (nonrestrictive, educational, etc.)? **OBJ 1.18**

Currency (CUR)

❑ Is the date the material was first created in any format included on the page? **CUR 1.1**

❑ Is the date the material was first placed on the server included on the page? **CUR 1.2**

❑ If the contents of the page have been revised, is the date (and time, if appropriate) the material was last revised included on the page? **CUR 1.3**

❑ To avoid confusion, are all dates in an internationally recognized format? Examples of dates in international format (dd mm yy) are 5 June 1997 and 21 January 1999. **CUR 1.4**

Coverage and Intended Audience (COV/IA)

❑ Is it clear what materials are included at the site? **COV/IA 1.1**

❑ If the page is still under construction, is the expected date of completion indicated? **COV/IA 1.2**

❑ Is the intended audience for the material clear? **COV/IA 2.1**

❑ If material is presented for several different audiences, is the intended audience for each type of material clear? **COV/IA 2.2**

Interaction and Transaction Features (INT/TRA)

❑ If any financial transactions occur at the site, does the site indicate what measures have been taken to ensure their security? **INT/TRA 1.1**

❑ If the business, organization, or person responsible for the page is requesting information from the user, is there a clear indication of how the information will be used? **INT/TRA 1.2**

❑ If cookies are used at the site, is the user notified? Is there an indication of what the cookies are used for and how long they last? **INT/TRA 1.3**

❑ Is there a feedback mechanism for users to comment about the site? **INT/TRA 1.5**

❑ Are any restrictions regarding downloading and other uses of the materials offered on the page clearly stated? **INT/TRA 1.9**

5

Keys to Information Quality in Advocacy Web Pages

```
┌─────────────────────────────────────────┐
│                                         │
│           Chapter Contents              │
│                                         │
│    •  Keys to Recognizing               │
│       an Advocacy Page                  │
│    •  Analysis of Advocacy Pages        │
│    •  The Advocacy Checklist: Keys      │
│       to Evaluating and Creating        │
│       Advocacy Web Pages                │
│                                         │
└─────────────────────────────────────────┘
```

KEYS TO RECOGNIZING AN ADVOCACY PAGE

An advocacy Web page is one with the primary purpose of influencing public opinion. The purpose may be either to influence people's ideas or to encourage activism, and either a single individual or a group of people may be responsible for the page.

Examples of advocacy organizations include the Democratic and Republican Parties, the National Right to Life Committee, and the National Abortion Rights Action League. The URL address of an advocacy page frequently ends in *.org* (organization) if the page is sponsored by a nonprofit organization.

An answer of "yes" to any of the following questions provides a good indication that the primary purpose of the page is advocacy. Does the page:

1. Seek to influence people's opinion on something?
2. Seek to influence the legislative process?
3. Encourage contributions of money?
4. Try to influence voters?
5. Promote a cause?
6. Attempt to increase membership in an organization?
7. Provide a point of contact for like-minded people?

ANALYSIS OF ADVOCACY PAGES

Figure 5.1 shows the home page of the Physicians for Social Responsibility (PSR). By looking at the page, we can identify the following characteristics that indicate the page is an advocacy page:

- The URL extension of the page, *.org*, indicates the page may be from a nonprofit organization.
- The goals of the organization are to promote "a world free of nuclear weapons, global environmental pollution, and gun violence."
- The page provides a point of contact for like-minded people.
- The page provides a link to a way of becoming a member of the organization.
- PSR encourages political activism by providing a link to a congressional directory for citizens to contact their representatives about the issues.

Figure 5.1 (as well as Fig. 5.2, another page from the PSR site) illustrates some of the elements that are important to include on advocacy pages.

The Advocacy Checklist provides a list of questions to consider when analyzing an advocacy page. Applying the general questions from the Checklist of Basic Elements and also the specific questions from the Advocacy Checklist can help a user determine:

- The nature of the advocacy organization responsible for the contents of the site or page.
- Whether the information on the page is likely to be reliable, authoritative, and trustworthy.
- Whether the information at the site is relevant to the user's information needs.

These same questions can also be used by a Web author to create advocacy pages that can be recognized as originating from a reliable source.

THE ADVOCACY CHECKLIST: KEYS TO EVALUATING AND CREATING ADVOCACY WEB PAGES

An advocacy Web page is one with the primary purpose of influencing public opinion. The following questions are intended to complement the general questions found on the Checklist of Basic Elements. The greater the number of questions on both the Checklist of Basic Elements and on the Advocacy Checklist answered "yes," the greater the likelihood that the quality of information on an advocacy Web page can be determined.

If the page you are analyzing is not a home page, it is important to return to the site's home page to answer the questions in the Authority of the Site's Home Page section of the checklist.

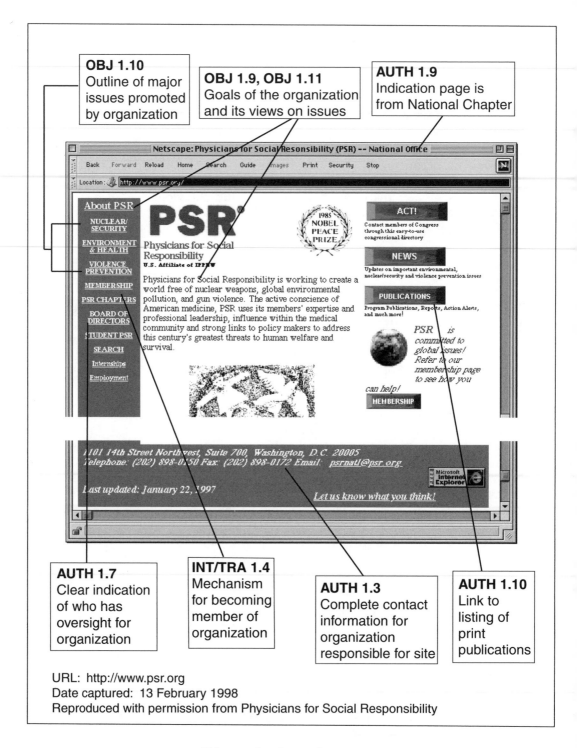

OBJ 1.10 Outline of major issues promoted by organization

OBJ 1.9, OBJ 1.11 Goals of the organization and its views on issues

AUTH 1.9 Indication page is from National Chapter

AUTH 1.7 Clear indication of who has oversight for organization

INT/TRA 1.4 Mechanism for becoming member of organization

AUTH 1.3 Complete contact information for organization responsible for site

AUTH 1.10 Link to listing of print publications

URL: http://www.psr.org
Date captured: 13 February 1998
Reproduced with permission from Physicians for Social Responsibility

FIG. 5.1. An advocacy home page.

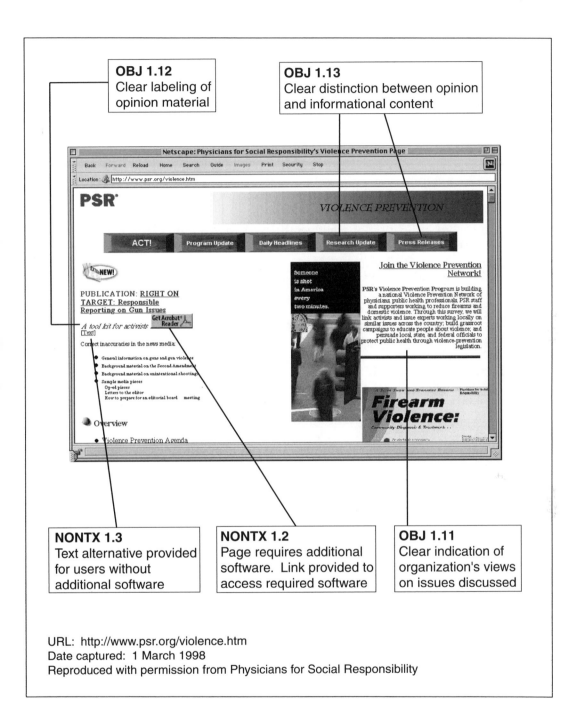

OBJ 1.12
Clear labeling of opinion material

OBJ 1.13
Clear distinction between opinion and informational content

NONTX 1.3
Text alternative provided for users without additional software

NONTX 1.2
Page requires additional software. Link provided to access required software

OBJ 1.11
Clear indication of organization's views on issues discussed

URL: http://www.psr.org/violence.htm
Date captured: 1 March 1998
Reproduced with permission from Physicians for Social Responsibility

FIG. 5.2. An advocacy Web page.

Authority

Authority of the Site's Home Page

The following information should be included either on the site's home page or on a page directly linked to the home page.

❑ Is there a listing of the names and qualifications of any individuals who are responsible for overseeing the organization (such as a Board of Directors)? **AUTH 1.7**

❑ Is there an indication of whether the advocacy organization has a presence beyond the Web? For example, do its members hold face-to-face meetings? **AUTH 1.8**

❑ Is there an indication whether the site is sponsored by an international, national, or local chapter of an organization? **AUTH 1.9**

❑ Is there a listing of printed materials produced by the organization and information about how they can be obtained? **AUTH 1.10**

❑ Is a complete description of the nature of the organization provided? **AUTH 1.11**

❑ Is there a statement of how long the organization has been in existence? **AUTH 1.12**

❑ Is there an indication that the organization adheres to guidelines established by an independent monitoring agency? **AUTH 1.14**

Objectivity

❑ Is there a description of the goals of the person or organization for providing the information? This is often found in a mission statement. **OBJ 1.9**

❑ Is it clear what issues are being promoted? **OBJ 1.10**

❑ Are the organization's or person's views on the issues clearly stated? **OBJ 1.11**

❑ Is there a clear distinction between expressions of opinion on a topic and any informational content that is intended to be objective? **OBJ 1.13**

Interaction and Transaction Features

❑ For sites with a membership option, is there a mechanism provided for users to become a member of the organization? **INT/TRA 1.4**

6

Keys to Information Quality in Business Web Pages

KEYS TO RECOGNIZING A BUSINESS PAGE

A business Web page is one with the primary purpose of promoting or selling products or services. Examples of uses for business Web pages include a store selling its products through an online catalog and a computer company providing upgrades for its software and other customer support services via the Web. The URL address of the page usually ends in *.com* (commercial).

An answer of "yes" to any of the following questions provides a good indication that the primary purpose of the page is business or marketing. Does the page:

1. Promote a product or service?
2. Provide customer support?
3. Make the company's catalog available online?
4. Provide product updates or new versions of a product?
5. Provide documentation about a product?
6. Request information about a person's lifestyle, demographics, or finances?

ANALYSIS OF BUSINESS PAGES

Figure 6.1 illustrates the home page from the Lands' End Direct Merchants Web site. The page provides an example of ways in which many important elements have been included on a business home page. Figures 6.2 and 6.3 are two additional pages from the Lands' End Web site that illustrate numerous other important features.

When analyzing a business Web page, it is important to first use the list of general questions found in the Checklist of Basic Elements, and subsequently apply the questions from the Business Checklist. Answering these questions can help a user determine:

- The nature of the business.
- Whether the information at the site is likely to be reliable, authoritative, and trustworthy.
- Whether the information at the site is relevant to a user's information needs.

These same questions can be used by Web authors as a guide to creating business pages that can be recognized as originating from a reliable, trustworthy source.

THE BUSINESS CHECKLIST: KEYS TO EVALUATING AND CREATING BUSINESS WEB PAGES

A business Web page is one with the primary purpose of promoting or selling products. The following questions are intended to complement the general questions found in the Checklist of Basic Elements. The greater the number of questions on both the Checklist of Basic Elements and the Business Checklist answered "yes," the greater the likelihood that the quality of information on a business Web page can be determined.

If the page you are analyzing is not a home page, it is important to return to the site's home page to answer the questions in the Authority of the Site's Home Page section of the checklist.

Authority

Authority of the Site's Home Page

The following information should be included either on the site's home page or on a page directly linked to the home page.

❏ Is it indicated whether the company has a presence beyond the Web? For example, does it indicate it has a printed catalog, or that it sells its merchandise in a traditional store? **AUTH 1.8**
❏ Is there a listing of printed materials about the company and its products and information about how they can be obtained? **AUTH 1.10**

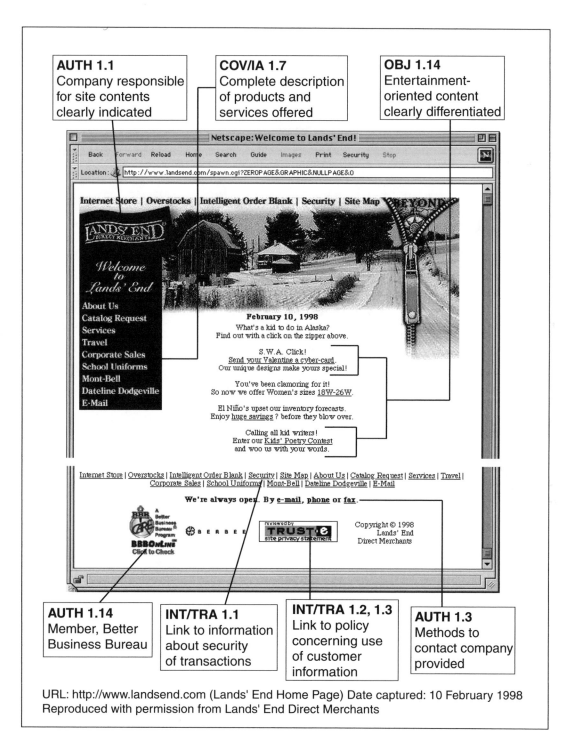

AUTH 1.1
Company responsible for site contents clearly indicated

COV/IA 1.7
Complete description of products and services offered

OBJ 1.14
Entertainment-oriented content clearly differentiated

AUTH 1.14
Member, Better Business Bureau

INT/TRA 1.1
Link to information about security of transactions

INT/TRA 1.2, 1.3
Link to policy concerning use of customer information

AUTH 1.3
Methods to contact company provided

URL: http://www.landsend.com (Lands' End Home Page) Date captured: 10 February 1998
Reproduced with permission from Lands' End Direct Merchants

FIG. 6.1. A business home page.

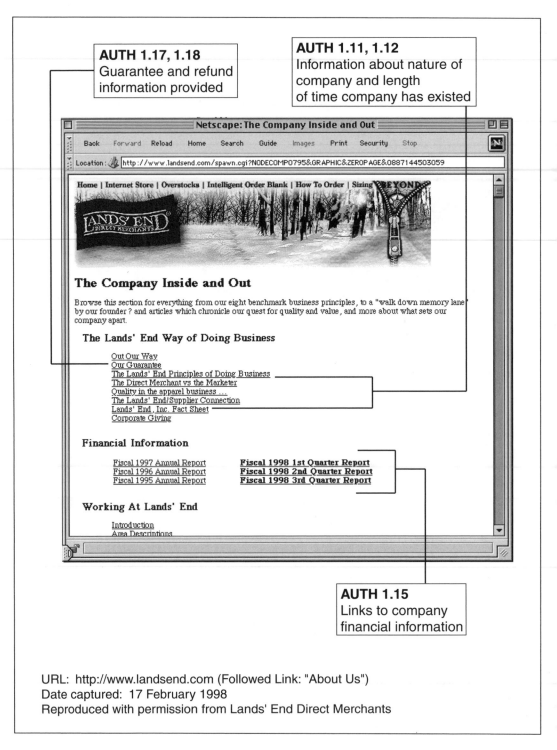

AUTH 1.17, 1.18
Guarantee and refund information provided

AUTH 1.11, 1.12
Information about nature of company and length of time company has existed

AUTH 1.15
Links to company financial information

URL: http://www.landsend.com (Followed Link: "About Us")
Date captured: 17 February 1998
Reproduced with permission from Lands' End Direct Merchants

FIG. 6.2. A business Web page.

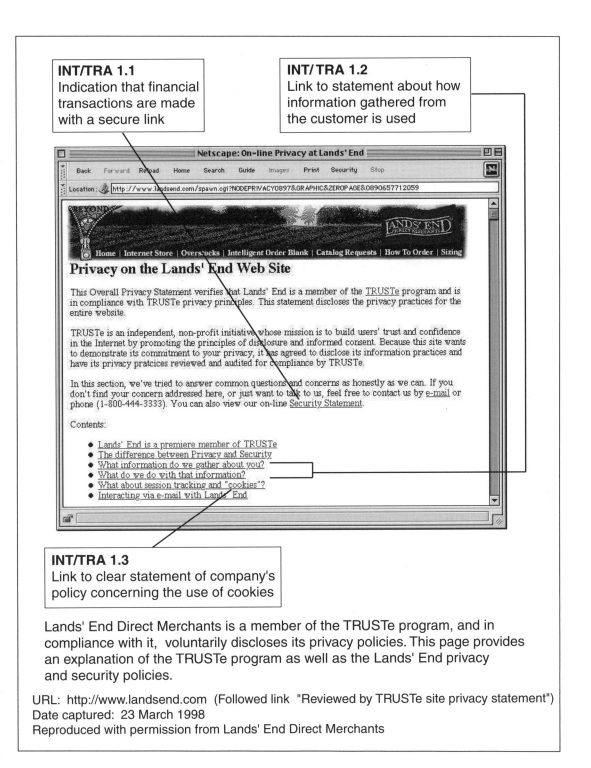

INT/TRA 1.1
Indication that financial transactions are made with a secure link

INT/ TRA 1.2
Link to statement about how information gathered from the customer is used

INT/TRA 1.3
Link to clear statement of company's policy concerning the use of cookies

Lands' End Direct Merchants is a member of the TRUSTe program, and in compliance with it, voluntarily discloses its privacy policies. This page provides an explanation of the TRUSTe program as well as the Lands' End privacy and security policies.

URL: http://www.landsend.com (Followed link "Reviewed by TRUSTe site privacy statement")
Date captured: 23 March 1998
Reproduced with permission from Lands' End Direct Merchants

FIG. 6.3. Explanation of site's privacy policy.

❑ Is a complete description of the nature of the company and the types of products or services offered provided? **AUTH 1.11**

❑ Is there a statement of how long the company has been in existence? **AUTH 1.12**

❑ Is there a listing of significant employees and their qualifications? **AUTH 1.13**

❑ Is there an indication that the company adheres to guidelines established by an independent monitoring agency such as the Better Business Bureau? **AUTH 1.14**

❑ Is company financial information provided? **AUTH 1.15**

❑ For financial information, is there an indication of whether it was filed with the Securities and Exchange Commission (SEC) and is a link provided to the SEC report? **AUTH 1.16**

❑ Is any warranty or guarantee information provided for the products or services of the business? **AUTH 1.17**

❑ Is there a refund policy indicated for any goods purchased from the site? **AUTH 1.18**

Accuracy

❑ Is there a link to outside sources such as product reviews or other independent evaluations of products or services that can be used to verify company claims? **ACC 1.6**

Objectivity

❑ If there is informational content not related to the company's products or services on the page, is it clear why the company is providing the information? **OBJ 1.8**

❑ If there is both information-oriented and entertainment-oriented content on the page, is there a clear differentiation between the two? **OBJ 1.14**

❑ If there is both advertising and entertainment-oriented content on the page, is there a clear differentiation between the two? **OBJ 1.15**

Currency

❑ If the page includes time-sensitive information, is the frequency of updates described? **CUR 1.5**

Coverage and Intended Audience

❑ Is there an adequately detailed description for the products and services offered? **COV/IA 1.7**

Interaction and Transaction Features

- ❑ Is there a mechanism for users to request additional information from the business and if so, is there an indication of when they will receive a response? **INT/TRA 1.6**
- ❑ Are there clear directions for placing an order for items available from the site? **INT/TRA 1.7**
- ❑ Is it clearly indicated when fees are required to access a portion of the site? **INT/TRA 1.8**

7

Keys to Information Quality in Informational Web Pages

KEYS TO RECOGNIZING AN INFORMATIONAL PAGE

An informational Web page is one with the primary purpose of providing factual information. Examples of materials found on informational pages include government research reports, census data, and factual information such as that found in encyclopedias. Information about a topic can be found on numerous different types of Web pages, so the URL address of an informational page may have any one of a variety of endings.

An answer of "yes" to any of the following questions provides a good indication that the primary purpose of the page is informational. Does the page provide:

1. Factual information about a topic?
2. Statistical information?
3. The results of research?
4. A schedule or calendar of events?
5. Transportation schedules?
7. Information such as that contained in a reference book?
8. A directory of names or businesses?
9. A list of course schedules?

70

ANALYSIS OF INFORMATIONAL PAGES

Figure 7.1 illustrates the home page of the Web site of WWW.CIOLEK.COM Asia Pacific Research Online. By looking at the page, we can identify the following characteristics:

- The URL extension of the page is a *.com*, indicating the page is probably from a commercial enterprise.
- Asia Pacific Research Online describes itself as a consulting organization in "networked knowledge management, in methods of high quality electronic publishing, and in Internet resources development and analysis."
- The page states that the "web server specializes in networked scholarly information related to online research, teaching and publishing about Asia and related regions and topics."
- The site accepts advertising and sponsorship.

These factors indicate that this site is a combination of a business site and an informational site. Figures 7.2 and 7.3 illustrate how some of the elements particularly important to include on a business home page and also an informational home page have been incorporated into the page.

For purposes of this chapter on informational Web pages, we have followed the "Online Information Resources" link to the Buddhist Studies WWW Virtual Library. Figures 7.2 and 7.3 illustrate the home page of the Buddhist Studies WWW Virtual Library, an informational subsite of the World Wide Web Virtual Library. By looking at the page's contents we can identify the following:

- "This document keeps track of leading information facilities in the fields of Buddhism and Buddhist Studies."
- "The page is edited by Dr. T. Matthew Ciolek and Professor Joe Brandsford Wilson in conjunction with other virtual librarians." A listing of qualifications for these editors is provided by means of a hypertext link.
- "Currently this and related pages provide direct WWW links to 320 specialist information facilities world-wide."

From these factors we can conclude that the page is, as the link indicates, an informational page. Figures 7.4 and 7.5 illustrate some of the factors it is important to include on a well-designed informational page.

Figures 7.6 and 7.7, from the Environmental Protection Agency (EPA) Web site, provide additional examples of how important elements have been incorporated into a site with primarily an informational focus. Figure 7.8, also from the EPA site, illustrates features important to include when presenting statistics on an informational page.

For additional examples of how important elements have been incorporated into the pages of an informational site, refer to Figs. 4.1 to 4.7 in chapter 4, which illustrate several pages from the OncoLink Web site.

URL: http://www.ciolek.com
Date captured: 12 March 1998
Reproduced with permission from Dr. T. Matthew Ciolek

FIG. 7.1. WWW.CIOLEK.COM Asia Pacific Research Online home page.

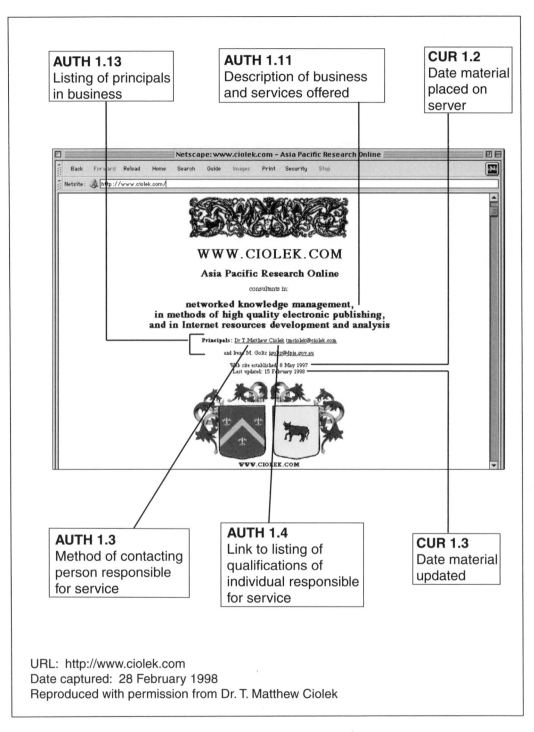

AUTH 1.13
Listing of principals in business

AUTH 1.11
Description of business and services offered

CUR 1.2
Date material placed on server

WWW.CIOLEK.COM

Asia Pacific Research Online

consultants in:

networked knowledge management, in methods of high quality electronic publishing, and in Internet resources development and analysis

Principals: Dr T.Matthew Ciolek tmciolek@ciolek.com
and Irena M. Goltz igoltz@dpie.gov.au

Web site established 8 May 1997
Last updated: 15 February 1998

VWV.CIOLEK.COM

AUTH 1.3
Method of contacting person responsible for service

AUTH 1.4
Link to listing of qualifications of individual responsible for service

CUR 1.3
Date material updated

URL: http://www.ciolek.com
Date captured: 28 February 1998
Reproduced with permission from Dr. T. Matthew Ciolek

FIG. 7.2. An informational and business home page (top). For bottom of page, see Fig. 7.3.

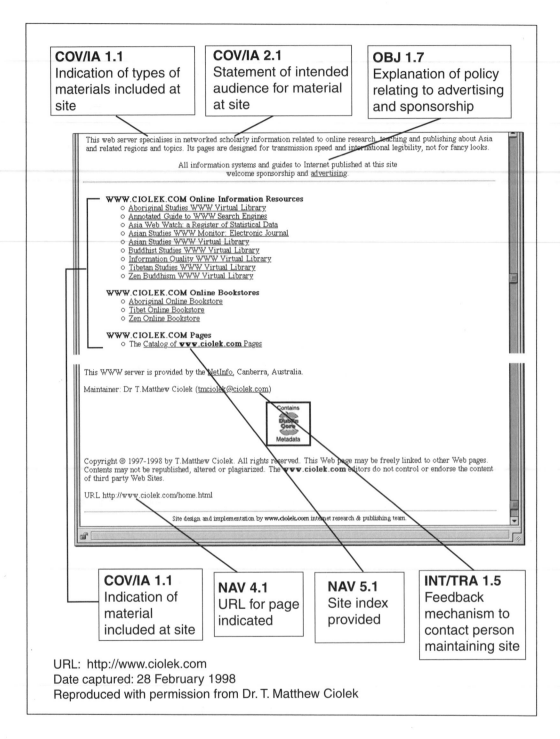

COV/IA 1.1
Indication of types of materials included at site

COV/IA 2.1
Statement of intended audience for material at site

OBJ 1.7
Explanation of policy relating to advertising and sponsorship

This web server specialises in networked scholarly information related to online research, teaching and publishing about Asia and related regions and topics. Its pages are designed for transmission speed and international legibility, not for fancy looks.

All information systems and guides to Internet published at this site welcome sponsorship and advertising.

WWW.CIOLEK.COM Online Information Resources
- Aboriginal Studies WWW Virtual Library
- Annotated Guide to WWW Search Engines
- Asia Web Watch: a Register of Statistical Data
- Asian Studies WWW Monitor: Electronic Journal
- Asian Studies WWW Virtual Library
- Buddhist Studies WWW Virtual Library
- Information Quality WWW Virtual Library
- Tibetan Studies WWW Virtual Library
- Zen Buddhism WWW Virtual Library

WWW.CIOLEK.COM Online Bookstores
- Aboriginal Online Bookstore
- Tibet Online Bookstore
- Zen Online Bookstore

WWW.CIOLEK.COM Pages
- The Catalog of **www.ciolek.com** Pages

This WWW server is provided by the NetInfo, Canberra, Australia.

Maintainer: Dr T.Matthew Ciolek (tmciolek@ciolek.com)

Copyright © 1997-1998 by T.Matthew Ciolek. All rights reserved. This Web page may be freely linked to other Web pages. Contents may not be republished, altered or plagiarized. The **www.ciolek.com** editors do not control or endorse the content of third party Web Sites.

URL http://www.ciolek.com/home.html

Site design and implementation by www.ciolek.com internet research & publishing team.

COV/IA 1.1
Indication of material included at site

NAV 4.1
URL for page indicated

NAV 5.1
Site index provided

INT/TRA 1.5
Feedback mechanism to contact person maintaining site

URL: http://www.ciolek.com
Date captured: 28 February 1998
Reproduced with permission from Dr. T. Matthew Ciolek

FIG. 7.3. An informational and business home page (bottom). For top of page, see Fig. 7.2.

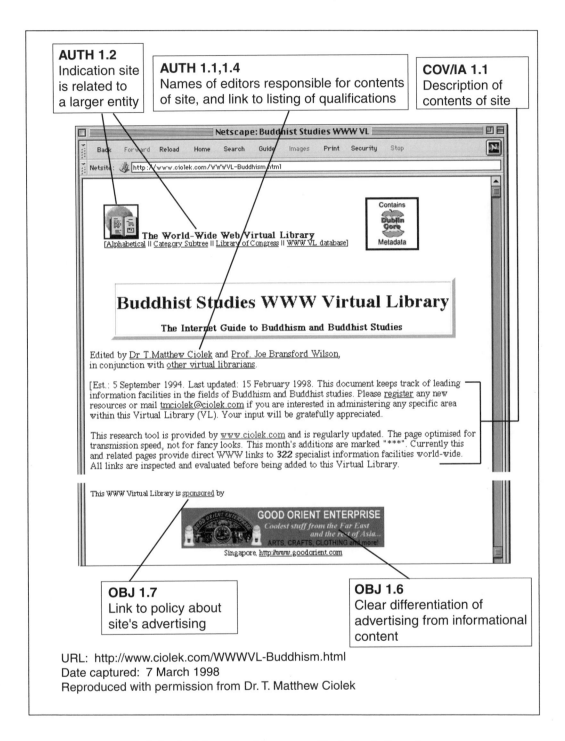

AUTH 1.2
Indication site is related to a larger entity

AUTH 1.1,1.4
Names of editors responsible for contents of site, and link to listing of qualifications

COV/IA 1.1
Description of contents of site

OBJ 1.7
Link to policy about site's advertising

OBJ 1.6
Clear differentiation of advertising from informational content

URL: http://www.ciolek.com/WWWVL-Buddhism.html
Date captured: 7 March 1998
Reproduced with permission from Dr. T. Matthew Ciolek

FIG. 7.4. An informational home page (top). For bottom of page, see Fig. 7.5.

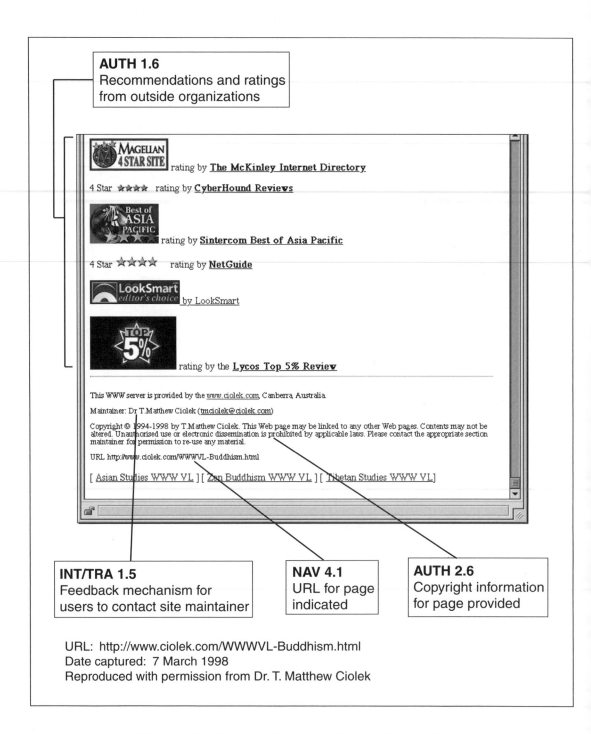

AUTH 1.6
Recommendations and ratings
from outside organizations

INT/TRA 1.5
Feedback mechanism for
users to contact site maintainer

NAV 4.1
URL for page
indicated

AUTH 2.6
Copyright information
for page provided

URL: http://www.ciolek.com/WWWVL-Buddhism.html
Date captured: 7 March 1998
Reproduced with permission from Dr. T. Matthew Ciolek

FIG. 7.5. An informational home page (bottom). For top of page,
see Fig. 7.4.

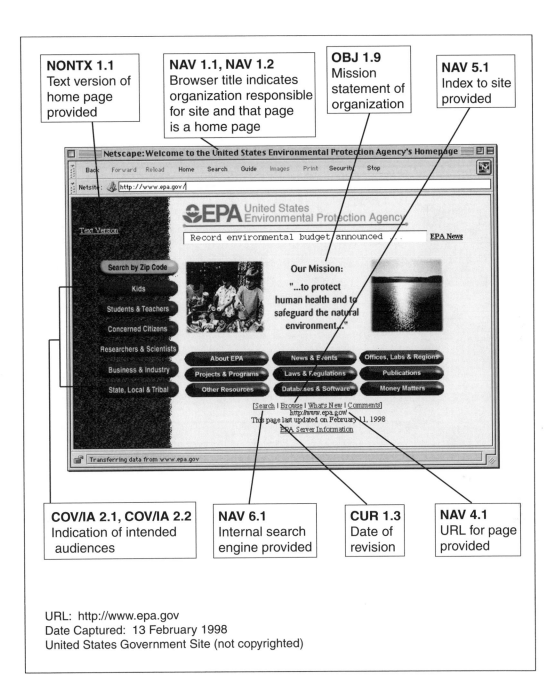

NONTX 1.1
Text version of home page provided

NAV 1.1, NAV 1.2
Browser title indicates organization responsible for site and that page is a home page

OBJ 1.9
Mission statement of organization

NAV 5.1
Index to site provided

COV/IA 2.1, COV/IA 2.2
Indication of intended audiences

NAV 6.1
Internal search engine provided

CUR 1.3
Date of revision

NAV 4.1
URL for page provided

URL: http://www.epa.gov
Date Captured: 13 February 1998
United States Government Site (not copyrighted)

FIG. 7.6. Additional example of an informational home page.

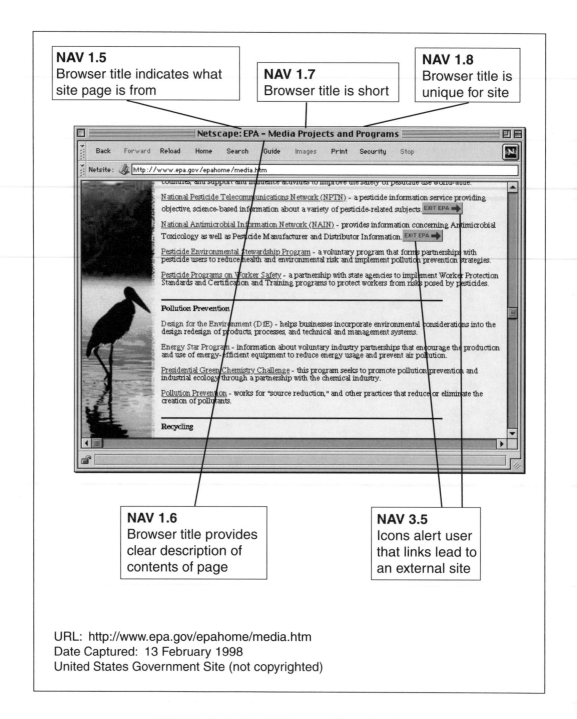

NAV 1.5
Browser title indicates what site page is from

NAV 1.7
Browser title is short

NAV 1.8
Browser title is unique for site

NAV 1.6
Browser title provides clear description of contents of page

NAV 3.5
Icons alert user that links lead to an external site

URL: http://www.epa.gov/epahome/media.htm
Date Captured: 13 February 1998
United States Government Site (not copyrighted)

FIG. 7.7. Navigational aids on an informational page.

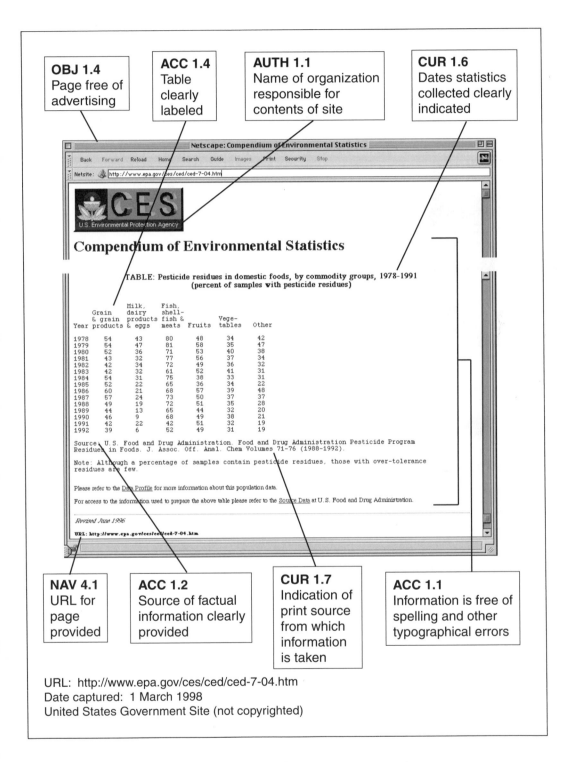

OBJ 1.4
Page free of advertising

ACC 1.4
Table clearly labeled

AUTH 1.1
Name of organization responsible for contents of site

CUR 1.6
Dates statistics collected clearly indicated

Netscape: Compendium of Environmental Statistics

Back   Forward   Reload   Home   Search   Guide   Images   Print   Security   Stop

Netsite: http://www.epa.gov/ces/ced/ced-7-04.htm

CES
U.S. Environmental Protection Agency

Compendium of Environmental Statistics

TABLE: Pesticide residues in domestic foods, by commodity groups, 1978-1991
(percent of samples with pesticide residues)

Year	Grain & grain products	Milk, dairy products & eggs	Fish, shell-fish & meats	Fruits	Vege-tables	Other
1978	54	43	80	48	34	42
1979	54	47	81	58	35	47
1980	52	36	71	53	40	38
1981	43	32	77	56	37	34
1982	42	34	72	49	36	32
1983	42	32	61	52	41	31
1984	54	31	75	38	33	31
1985	52	22	65	36	34	22
1986	60	21	68	57	39	48
1987	57	24	73	50	37	37
1988	49	19	72	51	35	28
1989	44	13	65	44	32	20
1990	46	9	68	49	38	21
1991	42	22	42	51	32	19
1992	39	6	52	49	31	19

Source: U.S. Food and Drug Administration. Food and Drug Administration Pesticide Program Residues in Foods. J. Assoc. Off. Anal. Chem Volumes 71-76 (1988-1992).

Note: Although a percentage of samples contain pesticide residues, those with over-tolerance residues are few.

Please refer to the Data Profile for more information about this population data.

For access to the information used to prepare the above table please refer to the Source Data at U.S. Food and Drug Administration.

Revised June 1996

URL: http://www.epa.gov/ces/ced/ced-7-04.htm

NAV 4.1
URL for page provided

ACC 1.2
Source of factual information clearly provided

CUR 1.7
Indication of print source from which information is taken

ACC 1.1
Information is free of spelling and other typographical errors

URL: http://www.epa.gov/ces/ced/ced-7-04.htm
Date captured: 1 March 1998
United States Government Site (not copyrighted)

FIG. 7.8. An informational page presenting statistics.

When analyzing an informational Web page, the first step is to ask the general questions listed in the Checklist of Basic Elements. In addition, a user must also apply the checklist questions from the Informational Checklist to determine:

- The nature of the information provider.
- Whether the information is likely to be reliable, authoritative, and trustworthy.
- Whether the information at the site is relevant to the user's information needs.

These same questions can be used by Web authors as a guide to creating informational Web pages that can be recognized as originating from a reliable, trustworthy source.

THE INFORMATIONAL CHECKLIST: KEYS TO EVALUATING AND CREATING INFORMATIONAL WEB PAGES

An informational Web page is one with the primary purpose of providing factual information. The following questions are intended to complement the general questions found in the Checklist of Basic Elements. The greater the number of questions on both the Checklist of Basic Elements and the Informational Checklist answered "yes,", the greater the likelihood that the quality of information on an informational Web page can be determined.

If the page you are analyzing is not a home page, it is important to return to the site's home page to answer the questions in the Authority of the Site's Home Page section of the checklist.

Authority

Authority of the Site's Home Page

The following information should be included either on the site's home page or on a page directly linked to the home page.

If an organization is responsible for the providing the information:

❑ Is there a listing of the names and qualifications of any individuals who are responsible for overseeing the organization (such as a Board of Directors)? **AUTH 1.7**
❑ Is there an indication of whether the organization has a presence beyond the Web? For example, does it produce printed materials? **AUTH 1.8**
❑ Is there a listing of printed materials produced by the organization and information about how they can be obtained? **AUTH 1.10**
❑ Is there a listing of significant employees and their qualifications? **AUTH 1.13**

Accuracy

- ❏ If the work is original research by the author, is this clearly indicated? **ACC 1.3**
- ❏ Is there an indication the information has been reviewed for accuracy by an editor or fact checker or through a peer review process? **ACC 1.5**

Currency

- ❏ If the page includes time-sensitive information, is the frequency of updates described? **CUR 1.5**
- ❏ If the page includes statistical data, is the date the statistics were collected clearly indicated? **CUR 1.6**
- ❏ If the same information is also published in a print source, such as an online dictionary with a print counterpart, is it clear which print edition the information is taken from? (Are the title, author, publisher, and date of the print publication listed?) **CUR 1.7**

Coverage and Intended Audience

- ❏ Is there is a print equivalent to the Web page? If so, is it clear if the entire work is available on the Web? **COV/IA 1.3**
- ❏ If there is a print equivalent to the Web page, is it clear if the Web version includes additional information not contained in the print version? **COV/IA 1.4**
- ❏ If the material is from a work that is out of copyright (as is often the case with a dictionary or thesaurus) is it clear whether and to what extent the material has been updated? **COV/IA 1.5**

8

Keys to Information Quality in News Web Pages

Chapter Contents

- **Keys to Recognizing a News Page**
- **Analysis of a News Pages**
- **The News Checklist: Keys to Evaluating and Creating News Web Pages**

KEYS TO RECOGNIZING A NEWS PAGE

A news Web page is one with the primary purpose of providing current information on local, regional, national, or international events. There are also numerous news sites devoted to one particular topic, such as business news, technology news, legal news, and so forth. The site may or may not have a print or broadcast equivalent. For organizations that have a non-Web counterpart, the Web version may or may not duplicate it.

Examples of some organizations with news Web sites include newspapers with a print counterpart, television and radio stations, and Web-based news organizations without a print counterpart. The URL address of a news page frequently ends in *.com* (commercial).

An answer of "yes" to either of the following questions provides a good indication that the primary purpose of the page you are analyzing is to provide news. Does the page:

- Provide current information on local, regional, national, or international events?
- Provide current information on a specific topic such as business, computers, or entertainment?

ANALYSIS OF NEWS PAGES

Figure 8.1 illustrates the home page of the *Washington Post/washingtonpost.com* Web site. Figures 8.2 and 8.3 are additional pages from the same site.

These pages provide examples of many of the elements that are important to include on news Web pages:

- The name of the organization responsible for the contents of the site.
- The date and time the page was last updated.
- A clear labeling of editorial and opinion material.
- An indication that the Web version includes additional information not contained in the print counterpart.
- A clear indication of which portions of the print equivalent are available on the Web site.
- The frequency of updates.
- An indication that the information provided has been reviewed by editors and fact checkers.

The questions in the News Checklist complement the general questions listed in the Checklist of Basic Elements. Applying the questions from both checklists to news Web page can assist a user in determining:

- Information about the authority of the news provider.
- The extent of news coverage provided at the site and how it differs from any non-Web counterpart.
- Whether the news provided at the site is relevant to the user's information needs.

THE NEWS CHECKLIST: KEYS TO EVALUATING AND CREATING NEWS WEB PAGES

A news Web page is one with the primary purpose of providing current information, either on local, regional, national or international events, or providing news in a particular subject area. The site may or may not have a print or broadcast equivalent. The following questions are intended to complement the general questions found in the Checklist of Basic Elements. The greater the number of questions on both the Checklist of Basic Elements and the News Checklist answered "yes," the greater the likelihood that the quality of information on a news Web page can be determined.

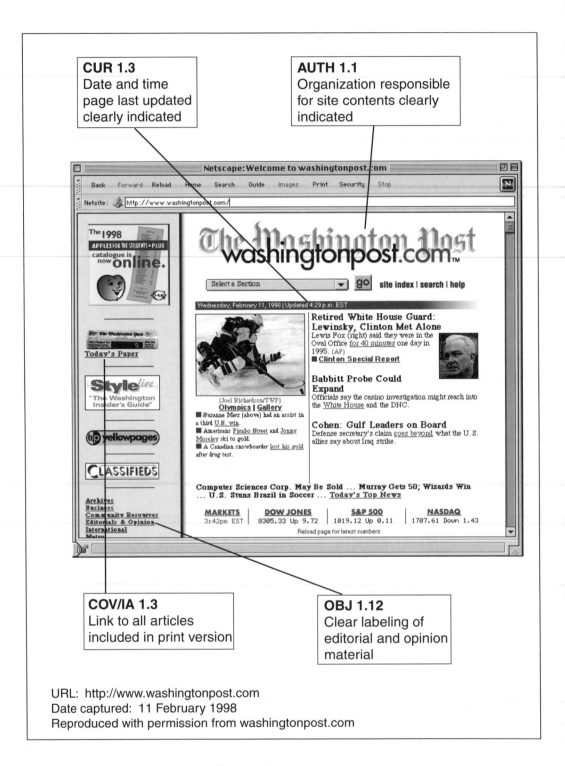

CUR 1.3
Date and time
page last updated
clearly indicated

AUTH 1.1
Organization responsible
for site contents clearly
indicated

COV/IA 1.3
Link to all articles
included in print version

OBJ 1.12
Clear labeling of
editorial and opinion
material

URL: http://www.washingtonpost.com
Date captured: 11 February 1998
Reproduced with permission from washingtonpost.com

FIG. 8.1. A news home page.

Netscape: WashingtonPost.com: User Guide

Back Forward Reload Home Search Guide Images Print Security Stop

Netsite: http://www.washingtonpost.com/wp-srv/guide/front.htm

washingtonpost.com | home page | site index | search | help |

About the Web Site

Washingtonpost.com is The Washington Post online — and much more. In addition to the full text of every story from the print edition, washingtonpost.com contains breaking news updates, special features and enhancements, a dynamic guide to arts and entertainment in the Washington area, searchable classifieds, our own interactive Yellow Pages and lively discussion areas.

The Post and washingtonpost.com
The Washington Post and washingtonpost.com are not exactly the same thing. Although we are owned by the same company and our missions are similar, washingtonpost.com has evolved to take advantage of the speed, depth and interactivity of this new medium.

That's why our Style or Metro section fronts look different from those in the paper. On washingtonpost.com, you'll see late-breaking stories from wire services, special reports from our own staff and dozens of databases about local communities, the 50 states and more than 200 countries. Washingtonpost.com has thousands of pages of information that never appeared in print, as well as all the stories published in the newspaper during the previous two weeks.

Our home page often resembles the newspaper's front page in the morning, then changes during the day as events warrant. Our news team adds stories to the news sections during the day and overnight. If you want, you can always see the full text of every story in the morning paper by clicking on the "Print Edition" link on the home page and most section fronts.

What's Online and What's Not
Every article in the newspaper is available on the Web site around midnight, and the final edition of the newspaper is posted by 6 a.m. The following material from The Post does not appear online: display advertising; some photos, graphics and tabular data; and syndicated comics and columns that washingtonpost.com does not have permission to reproduce.

For help finding a story, visit our main Search page, which is accessible by clicking on the word "Search" at the top of each page. There you can search the text of the paper, the full contents of our entertainment guide and Yellow Pages, and much more. You can even search the contents of The Washington Post going back to 1986 in the archives. Searching the Post archives is free and produces a list of story summaries, but retrieving the full text of a story costs between $1.50 and $2.95 each, depending on the time of day. The rest of the service is free. However, during a preview period, there is no charge for this service.

Site-Wide Navigation

URL: http://www.washingtonpost.com/wp-srv/guide/front.htm
Date captured: 11 February 1998
Reproduced with permission from washingtonpost.com

FIG. 8.2. Coverage of a news Web site.

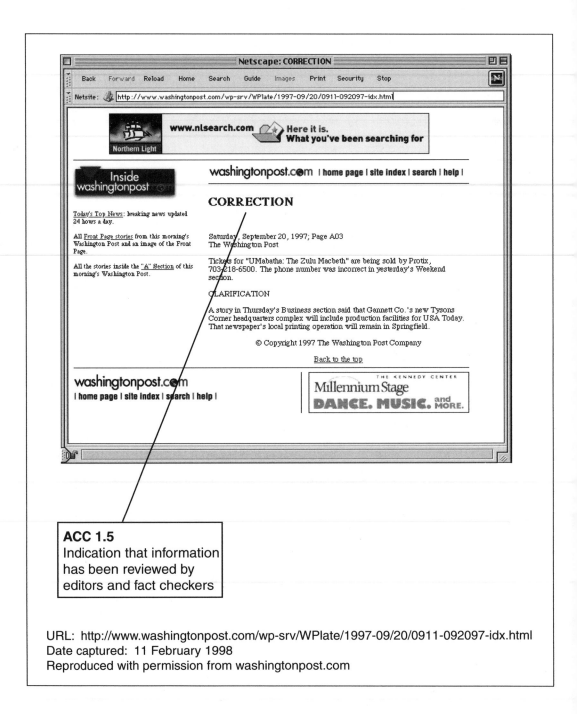

ACC 1.5
Indication that information
has been reviewed by
editors and fact checkers

URL: http://www.washingtonpost.com/wp-srv/WPlate/1997-09/20/0911-092097-idx.html
Date captured: 11 February 1998
Reproduced with permission from washingtonpost.com

FIG. 8.3. Example of corrections at a news Web site.

Authority

Authority of a Page Within the Site

❏ Is there a clear indication if the material has been taken from another source such as a newswire or news service? **AUTH 2.8**

Accuracy

❏ Is there an indication the information has been reviewed for accuracy by an editor or fact checker? **ACC 1.5**

Objectivity

❏ Is there a clear labeling of editorial and opinion material? **OBJ 1.12**

Currency

❏ If the page includes time-sensitive information, is the frequency of updates described? **CUR 1.5**

❏ If the same information also appears in print, is it clear which print edition the information is from (i.e., national, local, evening, morning edition, etc.)? **CUR 1.7**

❏ If the material was originally presented in broadcast form, is there a clear indication of the date and time the material was originally broadcast? **CUR 1.8**

Coverage and Intended Audience

❏ Is there is a print equivalent to the Web page or site? If so, is it clear if the entire work is available on the Web or if parts have been omitted? **COV/IA 1.3**

❏ If there is a print equivalent to the Web page, is it clear if the Web version includes additional information not contained in the print version? **COV/IA 1.4**

9

Keys to Information Quality in Personal Web Pages

┌─────────────────────────────────────┐
│   │
│   **Chapter Contents**   │
│   │
│ • **Keys to Recognizing**   │
│   **a Personal Page**   │
│ • **Analysis of Personal Pages**   │
│   │
└─────────────────────────────────────┘

KEYS TO RECOGNIZING A PERSONAL PAGE

A personal Web page is created by an individual who may or may not be affiliated with a larger institution. Personal pages are as diverse as the creators themselves and include, among numerous other things, pages that showcase an individual's artistic talents or are devoted to a favorite hobby or pastime. The URL address of a personal page may have a variety of endings depending on what type of site the page is coming from.

An answer of "yes" to any of the following questions provides a good indication that the page you are analyzing is a personal page. Does the page:

1. Have as its author a person or family with no official organizational affiliation?
2. Consist of a personal expression of something:
   • Hobbies or pastimes such as art, music or photography.
   • Personally authored plays, poems or other literature.
   • Personal opinions on a topic.

ANALYSIS OF PERSONAL PAGES

Figure 9.1 is an illustration of the home page of a personal Web site entitled "Mave's Media Haven." The inclusion of elements commonly found on other types of Web pages is a normal occurrence on personal Web pages. For example, the "Mave's Media Haven" home page includes links to:

- Advocacy organization pages.
- Informational pages.
- Entertainment pages.
- Business pages.

To analyze the various pages linked to this home page, it would be necessary to use the Checklist of Basic Elements as well as the appropriate individual checklists.

The "Who Is This Mave" link on the home page leads to information about the creator of the "Mave's Media Haven" site and other background information about the site. This type of information can help the user evaluate the authority and objectivity of the site and its creator.

Use the list of questions found in the Checklist of Basic Elements when analyzing a personal page. Applying the checklist questions to the "Mave's Media Haven" home page or to any other personal page can help determine:

- Who is responsible for the material on the page.
- Whether the material on the page is likely to be reliable, authoritative, and trustworthy.
- Whether the material at the site is relevant to the user's information needs.

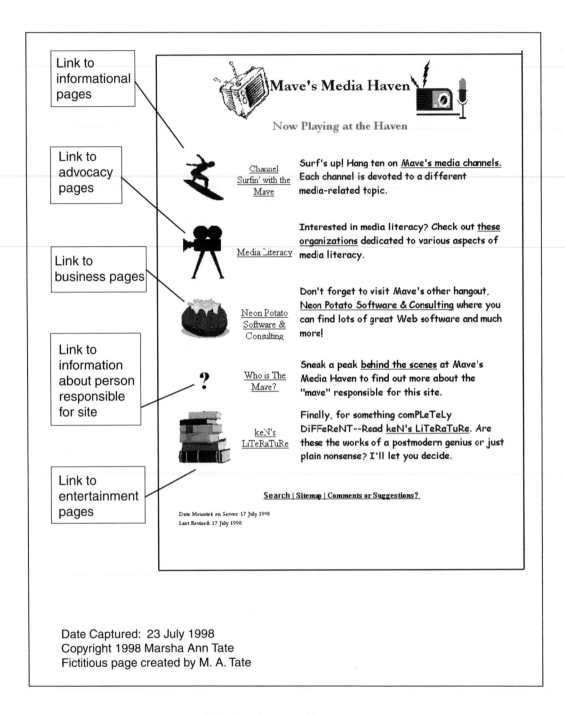

Link to informational pages

Link to advocacy pages

Link to business pages

Link to information about person responsible for site

Link to entertainment pages

Mave's Media Haven

Now Playing at the Haven

Channel Surfin' with the Mave

Surf's up! Hang ten on <u>Mave's media channels.</u> Each channel is devoted to a different media-related topic.

Media Literacy

Interested in media literacy? Check out <u>these organizations</u> dedicated to various aspects of media literacy.

Neon Potato Software & Consulting

Don't forget to visit Mave's other hangout, <u>Neon Potato Software & Consulting</u> where you can find lots of great Web software and much more!

Who is The Mave?

Sneak a peak <u>behind the scenes</u> at Mave's Media Haven to find out more about the "mave" responsible for this site.

keN's LiTeRaTuRe

Finally, for something comPLeTeLy DiFFeReNT--Read <u>keN's LiTeRaTuRe</u>. Are these the works of a postmodern genius or just plain nonsense? I'll let you decide.

<u>Search | Sitemap | Comments or Suggestions?</u>

Date Mounted on Server: 17 July 1998
Last Revised: 17 July 1998

Date Captured: 23 July 1998
Copyright 1998 Marsha Ann Tate
Fictitious page created by M. A. Tate

FIG. 9.1. A personal home page.

Keys to Information Quality in Entertainment Web Pages

+---+
| **Chapter Contents** |
| |
| • **Keys to Recognizing** |
| **an Entertainment Page** |
| • **Analysis of Entertainment Pages** |
| • **Entertainment Page Creation** |
| **Issues** |
+---+

KEYS TO RECOGNIZING AN ENTERTAINMENT PAGE

An entertainment Web page is one with the primary purpose of providing enjoyment to its users by means of humor, games, music, drama, or other similar types of activities. Examples of entertainment Web pages include pages that satirize other Web sites or pages that offer games, jokes, or fan fiction. The URL address of the page may have a variety of endings depending on who is supplying the entertainment.

An answer of "yes" to any of the following questions provides a good indication that the page is an entertainment page. Does the page:

1. Include games or other activities with the primary purpose of providing enjoyment?
2. Include music, animation, or video intended primarily to entertain its users?

Entertainment Pages: A Note for Web Page Users

If the primary purpose of the page is entertainment, enjoy the page. However, pages are not always created merely for entertainment but instead may also serve as a vehicle for business, marketing, or educational purposes. Examples of pages that perform these dual roles include ones that:

- Promote a product or service.
- Promote a company's public image.
- Teach an educational concept.
- Promote a TV or radio program or movie.

Such additional underlying purposes do not make the entertainment less enjoyable. We merely caution that Web users should, as they enjoy the entertainment, also be aware of these possible underlying motives and their potential influence on the user.

Knowing why entertainment is provided on a Web page becomes particularly important when children are targeted for marketing a product or for other promotional efforts. In addition, ascertaining the authority of an entertainment provider is also important when the payment of fees to the site is involved or when information is collected from a user, either openly via online registration forms and questionnaires, or transparently through the use of cookies. The procedure for evaluating a page that utilizes entertainment as a tool for promoting something else is similar to that already discussed earlier in the book:

- Use the checklist of Basic Elements to evaluate the page.
- Use additional checklists as appropriate.

For example, to evaluate Web pages that combine entertainment and product promotion, after first using the Checklist of Basic Elements, use the Business Checklist to analyze additional business aspects of the pages. In instances when entertainment is used to convey information, consult the informational checklist as well as the Checklist of Basic Elements.

ANALYSIS OF ENTERTAINMENT PAGES

Figures 10.1, 10.2, and 10.3 are examples of pages that use entertainment for a variety of purposes. Figure 10.1, a page from the Lands' End site, gives visitors an opportunity to send a virtual valentine, and is an example of entertainment provided by a company. Figure 10.2 is a Genway Products advertisement from the Minnesota Public Radio (MPR) site. Genway Products, a fictitious supermarket that stocks only "genetically engineered foods," is an example of entertainment used to promote an MPR radio show. Figure 10.3, a page from the Smithsonian Institution Web site, illustrates a page that uses entertainment for educational purposes.

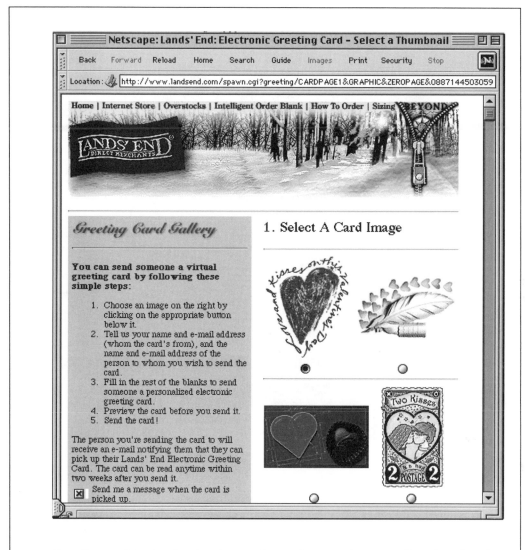

Sending a virtual valentine is one form of entertainment offered at the site. As well as providing entertainment, this type of service also promotes the Lands' End image.

URL: http://www.landsend.com (Followed link: "Send your Valentine a cyber-card" from home page)
Date captured: 10 February 1998
Reproduced with permission from Lands' End Direct Merchants

FIG. 10.1. Example of an entertainment page provided by a business.

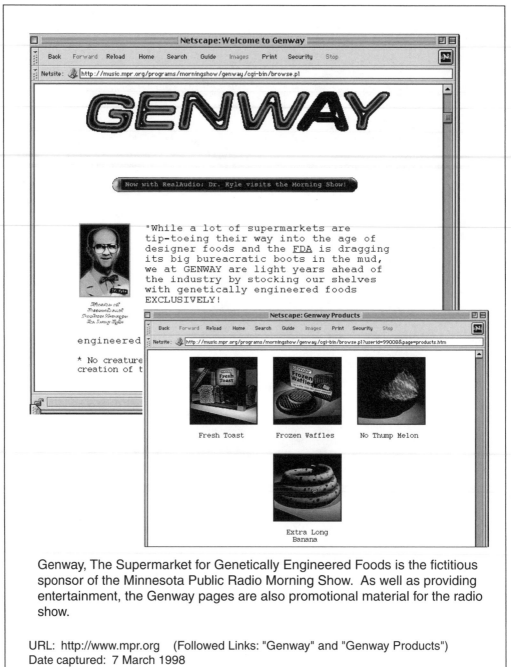

Genway, The Supermarket for Genetically Engineered Foods is the fictitious sponsor of the Minnesota Public Radio Morning Show. As well as providing entertainment, the Genway pages are also promotional material for the radio show.

URL: http://www.mpr.org   (Followed Links: "Genway" and "Genway Products")
Date captured: 7 March 1998
Reproduced with permission from Minnesota Public Radio

FIG. 10.2. Example of an entertainment page provided by a nonprofit corporation.

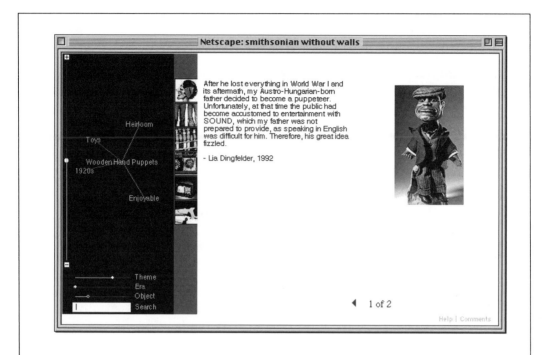

After he lost everything in World War I and its aftermath, my Austro-Hungarian-born father decided to become a puppeteer. Unfortunately, at that time the public had become accustomed to entertainment with SOUND, which my father was not prepared to provide, as speaking in English was difficult for him. Therefore, his great idea fizzled.

- Lia Dingfelder, 1992

This page is from the Smithsonian Institution's online exhibit "Revealing Things," an interactive display of objects from the museum's collection. This page presents historical information about wooden hand puppets in the United States. The entire exhibit presents, in an entertaining format, information that is primarily educational in nature.

URL: http://www.si.edu (From the exhibit "Revealing Things")
Date captured: 21 March 1998
Reproduced with permission from the Smithsonian Institution

FIG. 10.3. Example of a blending of entertainment and educational information.

Parody is another popular form of Web entertainment that often also serves a dual function. Figures 10.4 through 10.6 compare the U.S. government's official White House Web site with a parody of the site created by a business for promotional purposes. Figure 10.4 illustrates the U.S. government's official White House home page and Fig. 10.5 illustrates the parody page. In this instance, the parody White House page appears almost identical in design to the official White House page but with several somewhat obvious differences.

In addition to enjoying the humor of a parody site, it may also be useful to determine who created the site and for what purpose. An analysis of the White House parody home page (Fig. 10.6) reveals that:

- Some of the links are to pages at the official White House site.
- Some of the links are to parodies of other pages found at the official White House site.
- The link labeled "Why? Because we like you" is to a page describing the company responsible for the White House parody site.

By following the link "Why?" we find out that WHY? InterNetworking, a Web consultancy firm, is responsible for this particular White House site parody. WHY? InterNetworking is using the parody site to promote their services.

ENTERTAINMENT PAGE CREATION ISSUES

Creators of entertainment pages that have as their primary purpose providing enjoyment may not necessarily be concerned with authority, accuracy, currency, objectivity, and coverage issues. However, users should be given, at a minimum, information about who is providing the entertainment and who the intended audience is. In addition, depending on the type of entertainment offered, it may also be necessary for the creator to address other issues as well. The Checklist of Basic Elements can be used as a guide to help in the creation of an entertainment page.

FIG. 10.4. The U.S. government White House page.

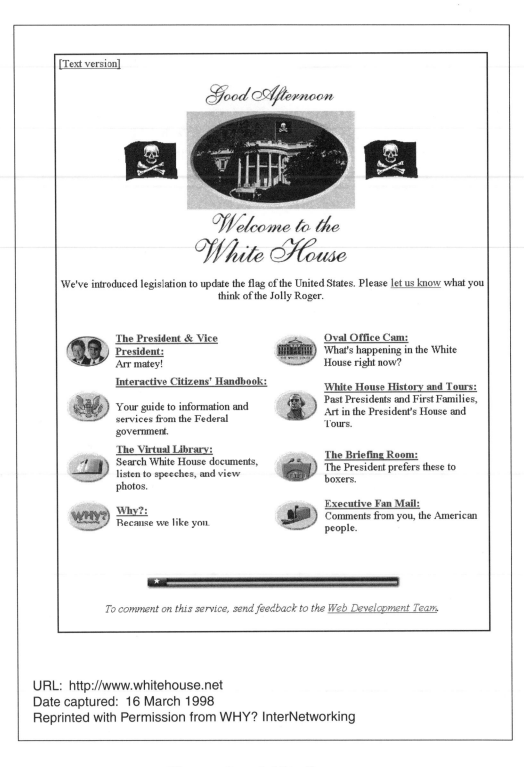

[Text version]

Good Afternoon

Welcome to the
White House

We've introduced legislation to update the flag of the United States. Please let us know what you think of the Jolly Roger.

The President & Vice President:
Arr matey!

Interactive Citizens' Handbook:
Your guide to information and services from the Federal government.

The Virtual Library:
Search White House documents, listen to speeches, and view photos.

Why?:
Because we like you.

Oval Office Cam:
What's happening in the White House right now?

White House History and Tours:
Past Presidents and First Families, Art in the President's House and Tours.

The Briefing Room:
The President prefers these to boxers.

Executive Fan Mail:
Comments from you, the American people.

To comment on this service, send feedback to the Web Development Team.

URL: http://www.whitehouse.net
Date captured: 16 March 1998
Reprinted with Permission from WHY? InterNetworking

FIG. 10.5. A parody White House page.

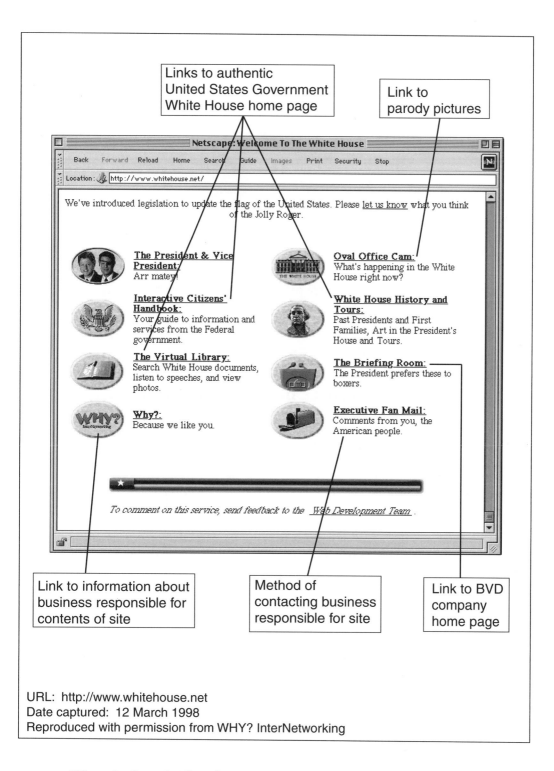

Links to authentic
United States Government
White House home page

Link to
parody pictures

Netscape: Welcome To The White House

Back Forward Reload Home Search Guide Images Print Security Stop

Location: http://www.whitehouse.net/

We've introduced legislation to update the flag of the United States. Please let us know what you think of the Jolly Roger.

The President & Vice President:
Arr matey!

Interactive Citizens' Handbook:
Your guide to information and services from the Federal government.

The Virtual Library:
Search White House documents, listen to speeches, and view photos.

Why?:
Because we like you.

Oval Office Cam:
What's happening in the White House right now?

White House History and Tours:
Past Presidents and First Families, Art in the President's House and Tours.

The Briefing Room:
The President prefers these to boxers.

Executive Fan Mail:
Comments from you, the American people.

To comment on this service, send feedback to the Web Development Team.

Link to information about
business responsible for
contents of site

Method of
contacting business
responsible for site

Link to BVD
company
home page

URL: http://www.whitehouse.net
Date captured: 12 March 1998
Reproduced with permission from WHY? InterNetworking

FIG. 10.6. A parody White House page (bottom). For entire page, see Fig. 10.5.

Creating Effective Web Pages and Sites

Chapter Contents

- **The Navigational Aids Checklist**
- **The Nontext Features Checklist**
- **The Interaction and Transaction Features Checklist**
- **The Web Site Functionality Checklist**
- **Meta Tags**
- **Copyright, Disclaimers, and Web Pages**

The first part of this chapter includes suggestions to help ensure that material at your site is accessible and easy to use. As stated in chapter 1, this book does not address visual design issues such as the use of graphics and color. It does, however, address design as it relates to the usability of a page. Any site must be easy enough to use that it does not frustrate its users or otherwise inhibit access to resources offered at the site. It does not matter how much care and attention has gone into creating a site of high information quality if the site is so poorly executed that people are deterred from using it. In addition, the chapter also addresses how to effectively facilitate interaction between Web users and your site. This chapter also addresses issues that help ensure that your site functions well. It concludes with a brief discussion of metatags and copyright issues.

This chapter provides checklists relating to the following features:

- Consistent and effective use of navigational aids.
- Appropriate use of nontext features such as graphics, frames, sound, and video.
- Effective handling of interaction and transaction features.
- Methods to help ensure your site functions well.

Following these suggestions will help assure that the pages you create will not confuse or frustrate users. For a basic explanation of navigational aids, nontext features, and interaction and transaction features, refer back to chapter 4.

The chapter also includes actual Web page examples (see Figs. 11.1–11.6) that further clarify how to effectively incorporate some of these features into Web pages and sites.

THE NAVIGATIONAL AIDS CHECKLIST

Navigational aids are elements that help a user locate information at a Web site and allow the user to easily move from page to page within the site. The greater the number of the following questions that are answered "yes," the more likely it is that the Web page you are creating has effective navigational aids.

NAV 1: Browser Titles

Browser Title for a Home Page

- ❑ Does the browser title indicate what company, organization, or person is responsible for the contents of the site? **NAV 1.1**
- ❑ Does the browser title indicate that the page is the main, or home page, for the site? **NAV 1.2**
- ❑ Is the browser title short? **NAV 1.3**
- ❑ Is the browser title unique for the site? **NAV 1.4**

Browser Title for Pages That Are Not Home Pages

- ❑ Does the browser title indicate what site the page is from? **NAV 1.5**
- ❑ Does the browser title clearly describe the contents of the page? **NAV 1.6**
- ❑ Is the browser title short? **NAV 1.7**
- ❑ Is the browser title unique for the site? **NAV 1.8**
- ❑ Does the browser title reflect the location of the page in the site hierarchy? **NAV 1.9**

NAV 2: The Page Title

Page Title for a Home Page

- ❑ Does the page title describe what site the page is from? This can be done using a logo. **NAV 2.1**
- ❑ Does the page title indicate that it is the main, or home, page for the site? **NAV 2.2**
- ❑ Is the page title short? **NAV 2.3**
- ❑ Is the page title unique for the site? **NAV 2.4**

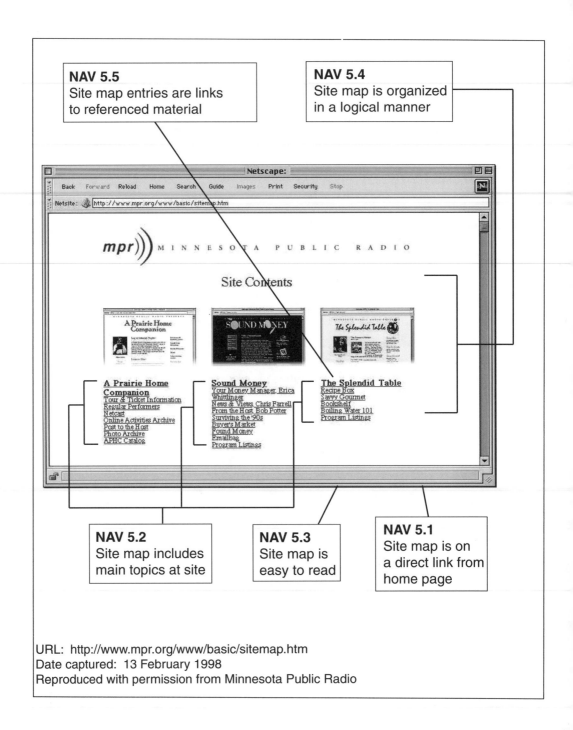

NAV 5.5
Site map entries are links
to referenced material

NAV 5.4
Site map is organized
in a logical manner

NAV 5.2
Site map includes
main topics at site

NAV 5.3
Site map is
easy to read

NAV 5.1
Site map is on
a direct link from
home page

URL: http://www.mpr.org/www/basic/sitemap.htm
Date captured: 13 February 1998
Reproduced with permission from Minnesota Public Radio

FIG. 11.1. Example of a site map.

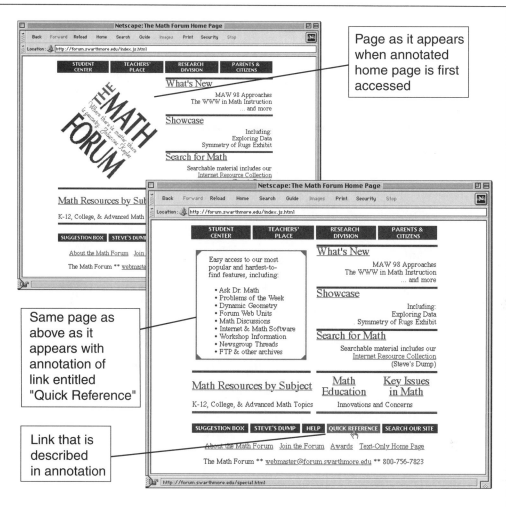

The Math Forum (a site designed to provide resources for math educators) provides an annotated home page in addition to its regular home page. The annotations serve a function similar to a more traditional site map, in that they indicate what material can be accessed by following a particular link. The annotations for "Quick Reference" appear when the cursor is placed over the "Quick Reference" link. The annotations are displayed using JavaScript.

URL: http://forum.swarthmore.edu/index.js.html
Date captured: 21 March 1998
Reproduced with permission from the Math Forum and Swarthmore College

FIG. 11.2. Example of a home page using annotated links to aid navigation.

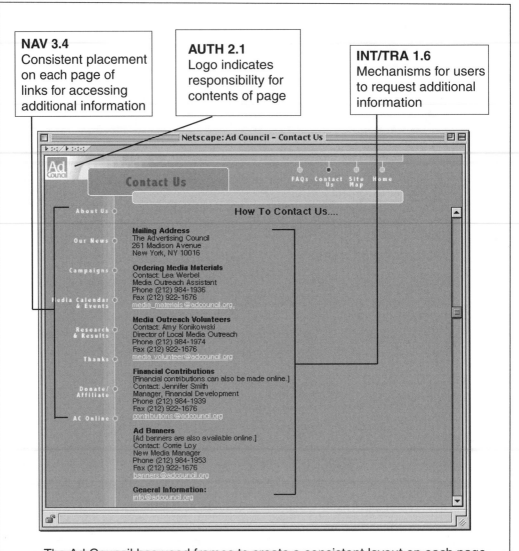

NAV 3.4
Consistent placement on each page of links for accessing additional information

AUTH 2.1
Logo indicates responsibility for contents of page

INT/TRA 1.6
Mechanisms for users to request additional information

The Ad Council has used frames to create a consistent layout on each page. This example also illustrates how separate contact information is provided for different types of requests for information.

URL: http://www.adcouncil.org/fr_contact.html
Date captured: 10 May 1998
Reproduced with permission from the Ad Council

FIG. 11.3. Example of a home page using a consistent layout and providing numerous contact points.

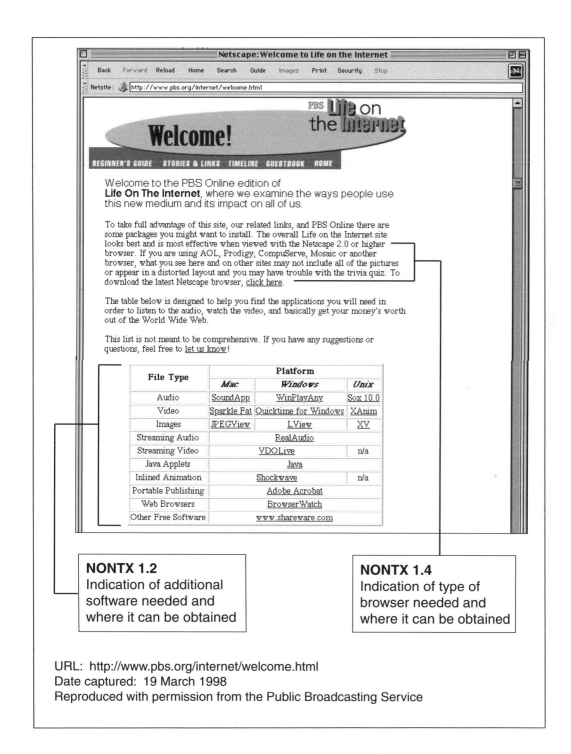

URL: http://www.pbs.org/internet/welcome.html
Date captured: 19 March 1998
Reproduced with permission from the Public Broadcasting Service

FIG. 11.4. Example of a site that indicates what software is needed to access information.

Netscape: Consumer Reports | Appliances

Back Forward Reload Home Search Guide Images Print Security Stop

Netsite: http://www.consumerreports.com/Categories/Appliances/index.html

Consumer Reports ONLINE   SITEMAP JOIN TALK SEARCH MORE HELP

CARS & TRUCKS
APPLIANCES
ELECTRONICS
HOUSE & HOME
HOME OFFICE
MONEY
HEALTH & FOOD
PERSONAL
LEISURE
MANUFACTURER LOCATOR
RECALLS

Appliances
Our brand-name Ratings & evaluations
Going shopping? Take along our reports, based on our own lab tests and readers' reliability experiences.

Safety alert

These two blenders can deliver a shock, we discovered

Ratings and recommendations
(Site subscriber area)

The following reports, from recent issues of Consumer Reports magazine, are available to site subscribers. **Click here to subscribe.** It costs $2.95 for a month's access to all of Consumer Reports Online.

Appliances quick-search: [Begin search]

INT/TRA 1.8
Clear indication fees are required to access portion of site

URL: http://www.consumerreports.com/Categories/Appliances/index.html
Date Captured: 7 March 1998
Copyright 1998 by Consumers Union of U.S., Inc., Yonkers, NY 10703-1057.
Reproduced by permission from CONSUMER REPORTS, March 1998

FIG. 11.5. Example of a site that indicates when a fee is required to access a portion of the site.

Copyright Statement:
"Copyright © 1997-1998 by T. Matthew Ciolek. All rights reserved. This Web page may be freely linked to other Web pages. Contents may not be republished, altered, or plagiarized. The **www.ciolek.com** editors do not control or endorse the content of third party Web Sites."

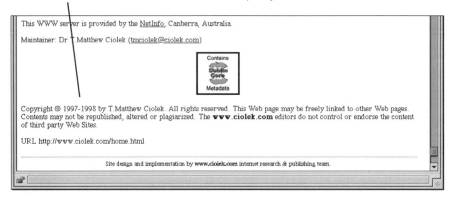

URL: http://www.ciolek.com/home.html
Date Captured: 28 February 1998
Reproduced with permission from Dr. T. Matthew Ciolek

Disclaimer Statement:
"OncoLink is designed for educational purposes only and is not engaged in rendering medical advice or professional services. The information provided through OncoLink should not be used for diagnosing or treating a health problem or a disease. It is not a substitute for professional care. If you have or suspect you may have a health problem, you should consult your health care provider."

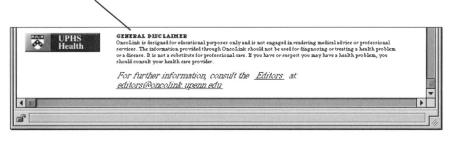

URL: http://www.oncolink.upenn.edu
Date Captured: 11 February 1998
Reproduced with permission from the University of Pennsylvania and OncoLink

FIG. 11.6. Example of copyright and disclaimer statements.

Page Title for a Page That Is Not a Home Page

❏ Does the page title clearly describe the contents of the page? **NAV 2.5**
❏ Is the page title short? **NAV 2.6**
❏ Is the page title unique for the site? **NAV 2.7**
❏ Does the page title give an indication of the company, organization, or person responsible for the contents of the site? **NAV 2.8**

NAV 3: Hypertext Links

❏ Does the page include a link to the home page? **NAV 3.1**
❏ Does the page include a link to a site map, index, or table of contents? **NAV 3.2**
❏ For sites arranged in a hierarchy, does the page include a link to the page one level up in the hierarchy? **NAV 3.3**
❏ Are internal directional links consistently placed on each page? **NAV 3.4**
❏ For links that access documents at an external site, is there an indication that the user will be leaving the site? **NAV 3.5**

NAV 4: The URL for the Page

❏ Does the page's URL appear in the body of the page? **NAV 4.1**

NAV 5: The Site Map or Index

❏ Is there a site map or index on the home page or on a page directly linked from the home page? **NAV 5.1**
❏ Does the site map include at a minimum the main topics at the site? **NAV 5.2**
❏ Is the site map or index easy to read? **NAV 5.3**
❏ Is the site map or index organized in a logical manner? **NAV 5.4**
❏ Are site map and index entries hypertext links to the referenced material? **NAV 5.5**

NAV 6: Internal Search Engine

❏ If your site provides large amounts of information, have you included an internal search engine at the site to enable users to locate the information quickly and easily? **NAV 6.1**
❏ Does the internal search engine retrieve complete and appropriate results? **NAV 6.2**

THE NONTEXT FEATURES CHECKLIST: KEYS TO EFFECTIVE USE

Nontext features include a wide array of elements that require the user to have additional software or a specific browser to utilize the contents of the page. Some examples of nontext features include image maps, sound, video, and graphics. The greater the number of the following questions that are answered "yes," the more likely the Web page you are creating is using nontext features appropriately.

Nontext Features (NONTX)

❑ If the page includes a graphic such as a logo or an image map, is there a text alternative for those viewing the page in text-only mode? **NONTX 1.1**

❑ If the page includes a nontext file (such as a sound or video file) that may require additional software, is there an indication of the additional software needed and where it can be obtained? **NONTX 1.2**

❑ If a file requires additional software to access it, wherever possible is the same information provided in another format that does not require the additional software? **NONTX 1.3**

❑ If a page requires a specific browser or a specific version of a browser, does the page specify what is needed and indicate where it can be obtained? **NONTX 1.4**

❑ When following a link results in the loading of a large graphic, sound, or video file, is information provided to alert the user that this will happen? **NONTX 1.5**

THE INTERACTION AND TRANSACTION FEATURES CHECKLIST: KEYS TO EFFECTIVE USE

Interaction features are mechanisms available at a Web site that enable a user to interact with the person or organization responsible for the site. Transaction features are tools that enable a user to enter into a transaction, usually financial, via the site. The greater the number of the following questions answered "yes," the more likely it is that your Web site deals appropriately with interaction and transaction features.

Interaction and Transaction Issues (INT/TRA)

❑ If any financial transactions occur at the site, does the site indicate what measures have been taken to ensure their security? **INT/TRA 1.1**

❑ If the business, organization, or person responsible for the site is requesting information from the user, is there a clear indication of how that information will be used? **INT/TRA 1.2**

❑ If cookies are used at the site, is the user notified? Is there an indication of what the cookies are used for and how long they last? **INT/TRA 1.3**

❑ For sites with a membership option, is there a mechanism provided for users to become a member of the organization? **INT/TRA 1.4**

❑ Is there a feedback mechanism for users to comment about the site? **INT/TRA 1.5**

❑ Is there a mechanism for users to request additional information from the organization or business and if so, is there an indication of when they will receive a response? **INT/TRA 1.6**

❑ Are there clear directions for placing an order for items available from the site? **INT/TRA 1.7**

❑ Is it clearly indicated when fees are required to access a portion of the site? **INT/TRA 1.8**

❑ Are any restrictions regarding downloading and other uses of the materials offered on the page clearly stated? **INT/TRA 1.9**

THE WEBSITE FUNCTIONALITY CHECKLIST: TESTING YOUR WEB SITE

Once your pages have been created, it is important to check them for accuracy and readability as well as a variety of other factors before you make them public. In addition, it is also important to check all links for functionality after the pages are placed on the server and periodically thereafter, to make certain that the links continue to function. The following are questions to ask to make sure that your Web site is functioning properly.

Printing Issues

❑ Have you checked to make sure pages print out legibly?

❑ Have any frames been tested to make sure that they can be printed out?

❑ If a long document has been divided into several different files, have you also made it possible to print out the same document in a single file?

Usability and Quality of External Links

❑ Do you test the functioning of external links when they are first added to your site?

❑ Do you test the functioning of external links on an ongoing basis to make sure they continue to link properly?

❑ Do you check the contents of external links on a regular schedule to make sure the links are still appropriate for your site and, if currency is an issue, have been kept up to date?

Usability of the Site

❑ Before making your pages public, have you tested them with people who will be using the site and modified them accordingly?

❑ Have you tested the pages to see how they look on as many different browsers as possible? (Whenever possible, create pages so they can be viewed correctly with as many browsers as possible.)

❑ Do you have a way of soliciting comments from the site's users on a regular basis concerning the layout and content of the site? Do you modify the site accordingly?

❑ Do you have an ongoing method for testing features at your site to make sure they are all functioning correctly? Features that need regular testing include:

- Internal links.
- External links.
- Forms.
- Images.
- Internal search engines.
- Animation.

❑ Do you remove outdated material on a regular basis?

❑ Do you indicate when new additions are placed on your site?

❑ Do you provide a method for accessing pages that have changed addresses?

❑ If major revisions have been made to a page, do you indicate what has been revised?

❑ For any printed documents at your site that have been converted to HTML and placed on your site, do you check to make sure the document has been converted completely and accurately?

❑ Do you provide an e-mail address for a "Webmaster" to whom people can write to inform you of any technical problems such as broken links?

META TAGS

A Brief Introduction

Several HTML tags called *meta tags* may allow Web page authors to exercise some control over

- How the page will be described when it appears in a list of results from a search engine query (*descriptor* meta tags).
- How search engines index the page (*keyword* meta tags).

The meta tags themselves will not be visible to the user viewing the Web page. For example, the information included within a descriptor meta tag is only visible when the Web page appears on a list of results from a search engine or when viewing the HTML source code.

Search engines vary widely in their treatment of meta tags, from using all of the meta information supplied by the page's author to ignoring the meta tags altogether.

However, failure to use meta tags would always place pages and sites at the mercy of whatever default formula the search engine uses to index and describe a web page.

All <META> tags are used within the <HEAD> element of a Web page

Descriptor Meta Tags

Descriptor meta tags allow Web page authors to provide a description of a Web page or site that can be used by a search engine when it retrieves the page as the result of a query. A descriptor meta tag can include up to 200 characters of text.

Failure to use the descriptor meta tag can result in a description of a Web page or site that gives a potential visitor either a poor idea of what they can expect to find at the page or site or, in some cases, no idea at all.

Example of a Descriptor Meta Tag

```
<HEAD>
<TITLE>Using Meta Tags When Creating Web Pages</TITLE>
<META name="description" content="This Web page describes how to
use meta tags."></HEAD>
```

The following title and description would appear if the page in the example were listed in the results for a search engine query:

Using Meta Tags When Creating Web Pages
This Web page describes how to use meta tags.

Keyword Meta Tags

A second important use of meta tags involves indexing terms. The Web includes a wide array of search engines, any number of which may index your Web page. However, the methods these search engines use to index pages vary greatly. Some search engines index all the words appearing on a Web page, whereas others index only portions of the page. However, just as the descriptor meta tag allows you to exert some control over how your page is described, the keywords meta tag allows you to supply some keywords that you think best characterize your site. The meta tag keywords will not be visible on your Web page but they can be used in the indexing process. The keywords meta tag can include up to 1,000 characters.

Tips for Using the Keyword Meta Tag

- Be sure that the keywords actually describe the materials available on your site.
- Use both common words and unique words (i.e., distinctive words that describe your site but few others).
- Use synonyms to supplement words included on your site.
- Provide full names for any important acronyms used on your page.

Example of a Keyword Meta Tag Included With a Descriptor Meta Tag

```
<HEAD>
<TITLE>Using Meta Tags When Creating Web Pages</TITLE>
<META name="description" content="This Web page describes how to
use the meta tag."
<META name="keywords" content="meta tags, keywords, Web page
creation">
</HEAD>
```

This page would be retrieved as the result of a search engine query that used either the words *meta tags*, *keywords*, or *web page creation* even though only one of these three terms (the words *meta tags*) is included in the actual text of the page itself.

COPYRIGHT, DISCLAIMERS, AND WEB PAGES

Copyright and the Web

The same factors that make the World Wide Web such a convenient channel of information exchange also raise many issues about copyright in the Web environment. Many of the questions raised have yet to be answered and these questions may not be fully resolved by the courts and legislative bodies for years to come. Therefore, with this in mind we can only offer some very general guidelines for Web authors to follow. It should also be noted that the following suggestions pertain to U.S. copyright law and therefore Web page creators outside the United States should consult the copyright laws for their country. In addition, Web authors in the United States are also strongly recommended to consult the U.S. Copyright Office Web site and other related resources to obtain further information and keep abreast of any changes in copyright law that are likely to occur in the near future. Finally, seek appropriate legal counsel if you need further advice and clarification on copyright matters.

Barron's *Law Dictionary* (Gifis, 1996) defines copyright as "the protection of the works of artists and authors giving them the exclusive right to publish their works or determine who may so publish" (p. 108).

Although copyright protection automatically begins the moment a Web page is created, there are several simple steps that you can take to ensure that you will be afforded maximum copyright protection. These steps include the use of copyright notices, copyright registration, and so on.

Works in the Public Domain (Works Not Protected by Copyright)

Copyright protection does not extend to all materials. Large numbers of works lack copyright protection. These materials include:

- Works the author has allowed to go into the public domain.
- Works for which the copyright has expired.
- Works that are authored by the federal government.

Although works in these categories may be used without prior permission, it is sometimes hard to determine whether a work falls within the public domain. When uncertainties arise, Copyright Office records can be searched to ascertain the current copyright status for a particular work.

Fair Use

The term *fair use* refers to a person's right to copy from a copyright-protected work for such purposes as criticism, comment, news reporting, teaching, scholarship, or research. According to the Copyright Act of 1976, Section 107, the following factors should be used to determine whether the use made of a work in any particular case is fair use:

1. The purpose and character of the use, including whether such use is of a commercial nature or is for nonprofit educated purposes.
2. The nature of the copyrighted work.
3. The amount and substantiality of the portion used in relation to the copyrighted work as a whole.
4. The effect of the use upon the potential market for or value of the copyrighted work. (Copyright Act of 1976, 17 U.S.C. Sect.107)

Fair use is yet another hotly debated issue in relation to the Web. Consult resources devoted to copyright issues for more information about the fair use concept as well as to learn of any possible changes to the fair use guidelines.

Copyright Notice

Although use of the copyright notice is not required to obtain copyright protection, it is still a good idea to place it on all of your Web pages. The notice serves as a visible sign to users of your materials that you have claimed ownership of the materials and the rights accompanying the ownership.

Copyright Notice Format

Use the following format when creating a copyright notice:

Copyright or ©   Publication date   Copyright owner's name

For Example:

Copyright   1998   Marsha Ann Tate

- Copyright vs. ©: Use the © whenever possible because in some countries the symbol, rather than the word *copyright*, represents the only legally acknowledged form of copyright. This is an especially important concern with the Web because Web materials have the potential for a worldwide audience.

- Publication date: The year in which the materials were first created.
- Copyright owner's name: Although there are various exceptions that allow individuals to use an alias in the copyright notice (if the person is identifiable by that alias), using your full name is probably the least problematic way to identify yourself.

Copyright Registration

Just as copyright notice is not a requirement for copyright protection, neither is registering your copyright with the Copyright Office. However, registration gives you a far greater opportunity to successfully defend your copyright ownership in any future disputes, as well as possibly recoup a larger portion of expenses you may incur in such a legal dispute. If you feel your material is important, take the time to register your copyright with the copyright office. Registration information and forms are available at the U.S. Copyright Office's Web site (http://lcweb.loc.gov/copyright/).

Suggested Copyright Guidelines for Web Authors

- Place your copyright notice on every Web page you create.
- Clearly state any additional restrictions you place on use of your materials (i.e., forbid usage of the materials without your "express permission," etc.).
- Make your copyright notice readable but nonobtrusive.
- Respect the copyright on any works you may include on your Web pages.

Search for your Web pages periodically on various search engines to monitor whether someone may be using your materials without your permission. This can be done by combining a search for the general topic of your page with several unusual words that appear on your page. If someone has borrowed your page without permission, the borrowed page may appear in the search results.

A Note on Disclaimers

If a site provides medical or any other type of information that may have potential liability issues, it would be wise to seek legal consultation to determine what type of disclaimer is appropriate for the site.

Appendix A

The Checklists:
A Compilation

THE CHECKLIST OF BASIC ELEMENTS: KEYS TO EVALUATING
OR CREATING WEB PAGES

The following questions are general ones that need to be asked when evaluating or creating any Web page, no matter what its type. Answering the following questions will help a user determine whether the information on a Web page is coming from an authoritative, accurate, and reliable source. The greater the number of "yes" answers, the greater the likelihood that the quality of the information on the page can be determined. The questions can also be used by Web authors as a guide to creating pages that can be recognized as originating from a reliable, trustworthy source.

Authority (AUTH)

Authority of a Site

The following information should be included either on a site's home page or on a page directly linked to it.

- ❑ Is it clear what organization, company, or person is responsible for the contents of the site? This can be indicated by the use of a logo. **AUTH 1.1**
- ❑ If the site is a subsite of a larger organization, does the site provide the logo or name of the larger organization? **AUTH 1.2**
- ❑ Is there a way to contact the organization, company, or person responsible for the contents of the site? These contact points can be used to verify the legitimacy of the site. Although a phone number, mailing address, and e-mail address are all possible contact points, a mailing address and phone number provide a more reliable way of verifying legitimacy. **AUTH 1.3**
- ❑ Are the qualifications of the organization, company, or person responsible for the contents of the site indicated? **AUTH 1.4**

❑ If all the materials on the site are protected by a single copyright holder, is the name of the copyright holder given? **AUTH 1.5**

❑ Does the site list any recommendations or ratings from outside sources? **AUTH 1.6**

Authority of a Page

❑ Is it clear what organization, company, or person is responsible for the contents of the page? Similarity in page layout and design features can help signify responsibility. **AUTH 2.1**

If the material on the page is written by an individual author:

> ❑ Is the author's name clearly indicated? **AUTH 2.2**
>
> ❑ Are the author's qualifications for providing the information stated? **AUTH 2.3**
>
> ❑ Is there a way of contacting the author? That is, does the person list a phone number, mailing address, and e-mail address? **AUTH 2.4**
>
> ❑ Is there a way of verifying the author's qualifications? That is, is there an indication of his or her expertise in the subject area, or a listing of memberships in professional organizations related to the topic? **AUTH 2.5**

❑ If the material on the page is copyright protected, is the name of the copyright holder given? **AUTH 2.6**

❑ Does the page have the official approval of the person, organization, or company responsible for the site? **AUTH 2.7**

Accuracy (ACC)

❑ Is the information free of grammatical, spelling, and typographical errors? **ACC 1.1**

❑ Are sources for factual information provided, so the facts can be verified in the original source? **ACC 1.2**

❑ If there are any graphs, charts, or tables, are they clearly labeled and easy to read? ACC 1.4

Objectivity (OBJ)

❑ Is the point of view of the individual or organization responsible for providing the information evident? **OBJ 1.1**

If there is an individual author of the material on the page:

> ❑ Is the point of view of the author evident? **OBJ 1.2**
>
> ❑ Is it clear what relationship exists between the author and the person, company, or organization responsible for the site? **OBJ 1.3**
>
> ❑ Is the page free of advertising? **OBJ 1.4**

For pages that include advertising:

  ❑ Is it clear what relationship exists between the business, organiza-
    tion, or person responsible for the contents of the page and any ad-
    vertisers represented on the page? **OBJ 1.5**
  ❑ If there is both advertising and information on the page, is there a
    clear differentiation between the two? **OBJ 1.6**
  ❑ Is there an explanation of the site's policy relating to advertising and
    sponsorship? **OBJ 1.7**

For pages that have a nonprofit or corporate sponsor:

  ❑ Are the names of any nonprofit or corporate sponsors clearly listed?
    OBJ 1.16
  ❑ Are links included to the sites of any nonprofit or corporate spon-
    sors so that a user may find out more information about them? **OBJ
    1.17**
  ❑ Is additional information provided about the nature of the sponsor-
    ship, such as what type it is (nonrestrictive, educational, etc.)? **OBJ
    1.18**

Currency (CUR)

  ❑ Is the date the material was first created in any format included on the
    page? **CUR 1.1**
  ❑ Is the date the material was first placed on the server included on the
    page? **CUR 1.2**
  ❑ If the contents of the page have been revised, is the date (and time, if ap-
    propriate) the material was last revised included on the page? **CUR 1.3**
  ❑ To avoid confusion, are all dates in an internationally recognized format?
    Examples of dates in international format (dd mm yy) are 5 June 1997
    and 21 January 1999. **CUR 1.4**

Coverage and Intended Audience (COV/IA)

  ❑ Is it clear what materials are included at the site? **COV/IA 1.1**
  ❑ If the page is still under construction, is the expected date of completion
    indicated? **COV/IA 1.2**
  ❑ Is the intended audience for the material clear? **COV/IA 2.1**
  ❑ If material is presented for several different audiences, is the intended au-
    dience for each type of material clear? **COV/IA 2.2**

Interaction and Transaction Features (INT/TRA)

  ❑ If any financial transactions occur at the site, does the site indicate what
    measures have been taken to ensure their security? **INT/TRA 1.1**

❏ If the business, organization, or person responsible for the page is requesting information from the user, is there a clear indication of how the information will be used? **INT/TRA 1.2**

❏ If cookies are used at the site, is the user notified? Is there an indication of what the cookies are used for and how long they last? **INT/TRA 1.3**

❏ Is there a feedback mechanism for users to comment about the site? **INT/TRA 1.5**

❏ Are any restrictions regarding downloading and other uses of the materials offered on the page clearly stated? **INT/TRA 1.9**

THE ADVOCACY CHECKLIST: KEYS TO EVALUATING AND CREATING ADVOCACY WEB PAGES

An advocacy Web page is one with the primary purpose of influencing public opinion. The following questions are intended to complement the general questions found on the Checklist of Basic Elements. The greater the number of questions on both the Checklist of Basic Elements and on the Advocacy Checklist answered "yes," the greater the likelihood that the quality of information on an advocacy Web page can be determined.

If the page you are analyzing is not a home page, it is important to return to the site's home page to answer the questions in the Authority of the Site's Home Page section of the checklist.

Authority

Authority of the Site's Home Page

The following information should be included either on the site's home page or on a page directly linked to the home page.

❏ Is there a listing of the names and qualifications of any individuals who are responsible for overseeing the organization (such as a Board of Directors)? **AUTH 1.7**

❏ Is there an indication of whether the advocacy organization has a presence beyond the Web? For example, do its members hold face-to-face meetings? **AUTH 1.8**

❏ Is there an indication whether the site is sponsored by an international, national, or local chapter of an organization? **AUTH 1.9**

❏ Is there a listing of printed materials produced by the organization and information about how they can be obtained? **AUTH 1.10**

❏ Is a complete description of the nature of the organization provided? **AUTH 1.11**

❏ Is there a statement of how long the organization has been in existence? **AUTH 1.12**

❏ Is there an indication that the organization adheres to guidelines established by an independent monitoring agency? **AUTH 1.14**

Objectivity

☐ Is there a description of the goals of the person or organization for providing the information? This is often found in a mission statement. **OBJ 1.9**
☐ Is it clear what issues are being promoted? **OBJ 1.10**
☐ Are the organization's or person's views on the issues clearly stated? **OBJ 1.11**
☐ Is there a clear distinction between expressions of opinion on a topic and any informational content that is intended to be objective? **OBJ 1.13**

Interaction and Transaction Features

☐ For sites with a membership option, is there a mechanism provided for users to become a member of the organization? **INT/TRA 1.4**

THE BUSINESS CHECKLIST: KEYS TO EVALUATING AND CREATING BUSINESS WEB PAGES

A business Web page is one with the primary purpose of promoting or selling products. The following questions are intended to complement the general questions found on the Checklist of Basic Elements. The greater the number of questions on both the Checklist of Basic Elements and the Business Checklist answered "yes," the greater the likelihood that the quality of information on a business Web page can be determined.

If the page you are analyzing is not a home page, it is important to return to the site's home page to answer the questions in the Authority of the Site's Home Page section of the checklist.

Authority

Authority of the Site's Home Page

The following information should be included either on the site's home page or on a page directly linked to the home page.

☐ Is it indicated whether the company has a presence beyond the Web? For example, does it indicate it has a printed catalog, or that it sells its merchandise in a traditional store? **AUTH 1.8**
☐ Is there a listing of printed materials about the company and its products and information about how they can be obtained? **AUTH 1.10**
☐ Is a complete description of the nature of the company and the types of products or services offered provided? **AUTH 1.11**
☐ Is there a statement of how long the company has been in existence? **AUTH 1.12**
☐ Is there a listing of significant employees and their qualifications? **AUTH 1.13**

❑ Is there an indication that the company adheres to guidelines established by an independent monitoring agency such as the Better Business Bureau? **AUTH 1.14**

❑ Is company financial information provided? **AUTH 1.15**

❑ For financial information, is there an indication of whether it was filed with the Securities and Exchange Commission (SEC) and is a link provided to the SEC report? **AUTH 1.16**

❑ Is any warranty or guarantee information provided for the products or services of the business? **AUTH 1.17**

❑ Is there a refund policy indicated for any goods purchased from the site? **AUTH 1.18**

Accuracy

❑ Is there a link to outside sources such as product reviews or other independent evaluations of products or services that can be used to verify company claims? **ACC 1.6**

Objectivity

❑ If there is informational content not related to the company's products or services on the page, is it clear why the company is providing the information? **OBJ 1.8**

❑ If there is both information-oriented and entertainment-oriented content on the page, is there a clear differentiation between the two? **OBJ 1.14**

❑ If there is both advertising and entertainment-oriented content on the page, is there a clear differentiation between the two? **OBJ 1.15**

Currency

❑ If the page includes time-sensitive information, is the frequency of updates described? **CUR 1.5**

Coverage and Intended Audience

❑ Is there an adequately detailed description for the products and services offered? **COV/IA 1.7**

Interaction and Transaction Features

❑ Is there a feedback mechanism for users to comment about the site? **INT/TRA 1.5**

❑ Is there a mechanism for users to request additional information from the business and if so, is there an indication of when they will receive a response? **INT/TRA 1.6**

❑ Are there clear directions for placing an order for items available from the site? **INT/TRA 1.7**

❑ Is it clearly indicated when fees are required to access a portion of the site? **INT/TRA 1.8**

THE INFORMATIONAL CHECKLIST: KEYS TO EVALUATING AND CREATING INFORMATIONAL WEB PAGES

An informational Web page is one with the primary purpose of providing factual information. The following questions are intended to complement the general questions found on the Checklist of Basic Elements. The greater the number of questions on both the Checklist of Basic Elements and the Informational Checklist answered "yes," the greater the likelihood that the quality of information on an informational Web page can be determined.

If the page you are analyzing is not a home page, it is important to return to the site's home page to answer the questions in the Authority of the Site's Home Page section of the checklist.

Authority

Authority of the Site's Home Page

The following information should be included either on the site's home page or on a page directly linked to the home page.

If an organization is responsible for the providing the information:

❑ Is there a listing of the names and qualifications of any individuals who are responsible for overseeing the organization (such as a Board of Directors)? **AUTH 1.7**

❑ Is there an indication of whether the organization has a presence beyond the Web? For example, does it produce printed materials? **AUTH 1.8**

❑ Is there a listing of printed materials produced by the organization and infomation about how they can be obtained? **AUTH 1.10**

❑ Is there a listing of significant employees and their qualifications? **AUTH 1.13**

Accuracy

❑ If the work is original research by the author, is this clearly indicated? **ACC 1.3**

❑ Is there an indication the information has been reviewed for accuracy by an editor or fact checker or through a peer review process? **ACC 1.5**

Currency

❑ If the page includes time-sensitive information, is the frequency of updates described? **CUR 1.5**

❑ If the page includes statistical data, is the date the statistics were collected clearly indicated? **CUR 1.6**

❑ If the same information is also published in a print source, such as an online dictionary with a print counterpart, is it clear which print edition the information is taken from? (Are the title, author, publisher, and date of the print publication listed?) **CUR 1.7**

Coverage and Intended Audience

❑ Is there is a print equivalent to the Web page? If so, is it clear if the entire work is available on the Web? **COV/IA 1.3**

❑ If there is a print equivalent to the Web page, is it clear if the Web version includes additional information not contained in the print version? **COV/IA 1.4**

❑ If the material is from a work that is out of copyright (as is often the case with a dictionary or thesaurus) is it clear whether and to what extent the material has been updated? **COV/IA 1.5**

THE NEWS CHECKLIST: KEYS TO EVALUATING AND CREATING NEWS WEB PAGES

A news Web page is one with the primary purpose of providing current information, either on local, regional, national or international events, or providing news in a particular subject area. The site may or may not have a print or broadcast equivalent.

The following questions are intended to complement the general questions found on the Checklist of Basic Elements. The greater the number of questions on both the Checklist of Basic Elements and the News Checklist answered "yes," the greater the likelihood that the quality of information on a news Web page can be determined.

Authority

Authority of a Page within the Site

❑ Is there a clear indication if the material has been taken from another source such as a newswire or news service? **AUTH 2.8**

Accuracy

❑ Is there an indication the information has been reviewed for accuracy by an editor or fact checker? **ACC 1.5**

Objectivity

❑ Is there a clear labeling of editorial and opinion material? **OBJ 1.12**

Currency

❑ If the page includes time-sensitive information, is the frequency of updates described? **CUR 1.5**

❑ If the same information also appears in print, is it clear which print edition the information is from (i.e., national, local, evening, morning edition, etc.)? **CUR 1.7**

❑ If the material was originally presented in broadcast form, is there a clear indication of the date and time the material was originally broadcast? **CUR 1.8**

Coverage and Intended Audience

❑ Is there is a print equivalent to the Web page or site? If so, is it clear if the entire work is available on the Web or if parts have been omitted? **COV/IA 1.3**

❑ If there is a print equivalent to the Web page, is it clear if the Web version includes additional information not contained in the print version? **COV/IA 1.4**

THE NAVIGATIONAL AIDS CHECKLIST: KEYS TO THEIR EFFECTIVE USE

Navigational aids are elements that help a user locate information at a Web site and easily move from page to page within the site. The greater the number of the following questions are answered "yes," the more likely it is that the Web page you are creating has effective navigational aids.

NAV 1: Browser Titles

Browser Title for a Home Page

❑ Does the browser title indicate what company, organization, or person is responsible for the contents of the site? **NAV 1.1**

❑ Does the browser title indicate that the page is the main, or home page, for the site? **NAV 1.2**

❑ Is the browser title short? **NAV 1.3**

❑ Is the browser title unique for the site? **NAV 1.4**

Browser Title for Pages That Are Not Home Pages

❑ Does the browser title indicate what site the page is from? **NAV 1.5**

❑ Does the browser title clearly describe the contents of the page? **NAV 1.6**

❑ Is the browser title short? **NAV 1.7**
❑ Is the browser title unique for the site? **NAV 1.8**
❑ Does the browser title reflect the location of the page in the site's hierarchy? **NAV 1.9**

NAV 2: The Page Title

Page Title for a Home Page

❑ Does the page title describe what site the page is from? This can be done using a logo. **NAV 2.1**
❑ Does the page title indicate that it is the main, or home, page for the site? **NAV 2.2**
❑ Is the page title short? **NAV 2.3**
❑ Is the page title unique for the site? **NAV 2.4**

Page Title for a Page That is Not a Home Page

❑ Does the page title clearly describe the contents of the page? **NAV 2.5**
❑ Is the page title short? **NAV 2.6**
❑ Is the page title unique for the site? **NAV 2.7**
❑ Does the page title give an indication of the company, organization, or person responsible for the contents of the site? **NAV 2.8**

NAV 3: Hypertext Links

❑ Does the page include a link to the home page? **NAV 3.1**
❑ Does the page include a link to a site map, index, or table of contents? **NAV 3.2**
❑ For sites arranged in a hierarchy, does the page include a link to the page one level up in the hierarchy? **NAV 3.3**
❑ Are internal directional links consistently placed on each page? **NAV 3.4**
❑ For links that access documents at an external site, is there an indication that the user will be leaving the site? **NAV 3.5**

NAV 4: The URL for the Page

❑ Does the page's URL appear in the body of the page? **NAV 4.1**

NAV 5: The Site Map or Index

❑ Is there a site map or index on the home page or on a page directly linked from the home page? **NAV 5.1**
❑ Does the site map include at a minimum the main topics at the site? **NAV 5.2**

❑ Is the site map or index easy to read? **NAV 5.3**
❑ Is the site map or index organized in a logical manner? **NAV 5.4**
❑ Are site map and index entries hypertext links to the referenced material?
   NAV 5.5

NAV 6: Internal Search Engine

❑ If the site provides large amounts of information, does it include an internal search engine to enable users to locate the information quickly and easily? **NAV 6.1**
❑ Does the internal search engine retrieve complete and appropriate results? **NAV 6.2**

THE NONTEXT FEATURES CHECKLIST: KEYS TO EFFECTIVE USE

Nontext features include a wide array of elements that require the user to have additional software or a specific browser to utilize the contents of the page. Some examples of nontext features include image maps, sound, video, and graphics.

The greater the number of the following questions that are answered "yes," the more likely the Web page you are creating is using nontext features appropriately.

Nontext Features (NONTX)

❑ If the page includes a graphic such as a logo or an image map, is there a text alternative for those viewing the page in text-only mode? **NONTX 1.1**
❑ If the page includes a nontext file (such as a sound or video file) that may require additional software, is there an indication of the additional software needed and where it can be obtained? **NONTX 1.2**
❑ If a file requires additional software to access it, wherever possible is the same information provided in another format that does not require the additional software? **NONTX 1.3**
❑ If a page requires a specific browser or a specific version of a browser does the page specify what is needed and indicate where it can be obtained? **NONTX 1.4**
❑ When following a link results in the loading of a large graphic, sound, or video file, is information provided to alert the user that this will happen? **NONTX 1.5**

THE INTERACTION AND TRANSACTION FEATURES CHECKLIST: KEYS TO EFFECTIVE USE

Interaction features are feedback mechanisms available at a Web site that enable a user to interact with the person or organization responsible for the site. Transaction features are tools that enable a user to enter into a transaction, usually financial, with the site. The greater the number of the following questions answered "yes," the more likely it is that your Web site deals appropriately with interaction and transactions features.

Interaction and Transaction Features (INT/TRA)

❑ If any financial transactions occur at the site, does the site indicate what measures have been taken to ensure their security? **INT/TRA 1.1**

❑ If the business, organization, or person responsible for the site is requesting information from the user, is there a clear indication of how that information will be used? **INT/TRA 1.2**

❑ If cookies are used at the site, is the user notified? Is there an indication of what the cookies are used for and how long they last? **INT/TRA 1.3**

❑ For sites with a membership option, is there a mechanism provided for users to become a member of the organization? **INT/TRA 1.4**

❑ Is there a feedback mechanism for users to comment about the site? **INT/TRA 1.5**

❑ Is there a mechanism for users to request additional information from the organization or business and if so, is there an indication of when they will receive a response? **INT/TRA 1.6**

❑ Are there clear directions for placing an order for items available from the site? **INT/TRA 1.7**

❑ Is it clearly indicated when fees are required to access a portion of the site? **INT/TRA 1.8**

❑ Are any restrictions regarding downloading and other uses of the materials offered on the page clearly stated? **INT/TRA 1.9**

Appendix B

Information Quality Questions:
A Compilation

<table>
<tr><td>

Chapter Contents

- **Definitions of the Eight Major Categories of Information Quality Elements**
- **A Complete Listing by Category of Questions to Consider When Evaluating or Creating Web Pages**
- **Unique Identifiers for Each Question**

</td></tr>
</table>

AUTHORITY (AUTH)

Definition: The extent to which material is the creation of a person or organization that is recognized as having definitive knowledge of a given subject area.

Questions to Ask About a Site's Home Page

AUTH 1.1 Is it clear what organization, company, or person is responsible for the contents of the site? This can be indicated by the use of a logo.

AUTH 1.2 If the site is a subsite of a larger organization, does the site provide the logo or name of the larger organization?

AUTH 1.3 Is there a way to contact the organization, company, or person responsible for the contents of the site? These contact points can be used to verify the legitimacy of the site. Although a phone number, mailing address, and e-mail address are all possible contact points, a mailing address and phone number provide a more reliable way of verifying legitimacy.

AUTH 1.4 Are the qualifications of the organization, company, or person responsible for the contents of the site indicated?

AUTH 1.5   If all the materials on the site are protected by a single copyright holder, is the name of the copyright holder given?

AUTH 1.6   Does the site list any recommendations or ratings from outside organizations?

AUTH 1.7   Is there a listing of the names and qualifications of any individuals who are responsible for overseeing the organization (such as a Board of Directors?)

AUTH 1.8   Is there an indication of whether the organization or company has a presence beyond the Web? For example: does it hold face to face meetings, produce printed products, or have a traditional store?

AUTH 1.9   Is there an indication whether the site is sponsored by an international, national, or local chapter of an organization?

AUTH 1.10   Is there a listing of printed materials produced by the organization and information about how they can be obtained?

AUTH 1.11   Is a complete description of the nature of the company or organization provided?

AUTH 1.12   Is there a statement of how long the company or organization has been in existence?

AUTH 1.13   Is there a listing of significant employees and their qualifications?

AUTH 1.14   Is there an indication that the company or organization adheres to guidelines established by an independent monitoring agency?

AUTH 1.15   Is company financial information provided?

AUTH 1.16   For financial information, is there an indication of whether it was filed with the Securities and Exchange Commission (SEC) and is a link provided to the SEC report?

AUTH 1.17   Is any warranty or guarantee information provided for the products and services offered?

AUTH 1.18   Is there a refund policy indicated for any goods purchased from the site?

Questions to Ask About a Page That Is Not a Home Page

AUTH 2.1   Is it clear what organization, company, or person is responsible for the contents of the page? Similarity in page layout and design features can help signify responsibility.

If the material on the page is written by an individual author:

AUTH 2.2   Is the author's name clearly indicated?

AUTH 2.3   Are the author's qualifications for providing the information stated?

AUTH 2.4   Is there a way of contacting the author? That is, does the person list a phone number, mailing address, and e-mail address?

AUTH 2.5 Is there a way of verifying the author's qualifications? That is, is there an indication of his or her expertise in the subject area, or a listing of memberships in professional organizations related to the topic?

AUTH 2.6 If the material on the page is copyright protected, is the name of the copyright holder given?

AUTH 2.7 Does the page have the official approval of the person, organization, or company responsible for the site?

AUTH 2.8 Is there a clear indication if the material has been taken from another source such as a newswire or news service?

ACCURACY (ACC)

Definition: The extent to which information is reliable and free from errors.

Questions to Ask

ACC 1.1 Is the information free of grammatical, spelling, and typographical errors?

ACC 1.2 Are sources for factual information provided, so the facts can be verified in the original source?

ACC 1.3 If the work is original research by the author, is this clearly indicated?

ACC 1.4 If there are graphs, charts, or tables, are they clearly labeled and easy to read?

ACC 1.5 Is there an indication the information has been reviewed for accuracy by an editor or fact checker, or through a peer review process?

ACC 1.6 Is there a link to outside sources such as product reviews or other independent evaluations of products or services that can be used to verify company claims?

OBJECTIVITY (OBJ)

Definition: The extent to which material expresses facts or information without distortion by personal feelings or other biases.

Questions to Ask

OBJ 1.1 Is the point of view of the individual or organization responsible for providing the information evident?

If there is an individual author to the material on the page:

OBJ 1.2 Is the point of view of the author evident?

OBJ 1.3   Is it clear what relationship exists between the author and the person, company, or organization responsible for the site?

OBJ 1.4   Is the page free of advertising?

For pages that include advertising:

OBJ 1.5   Is it clear what relationship exists between the business, organization, or person responsible for the contents of the page and any advertisers represented on the page?

OBJ 1.6   If there is both advertising and information on the page, is there a clear differentiation between the two?

OBJ 1.7   Is there an explanation of the site's policy relating to advertising and sponsorship?

OBJ 1.8   If there is informational content not related to the company's products or services on the page, is it clear why the company is providing the information?

OBJ 1.9   Is there a description of the goals of the person or organization for providing the information? This is often found in a mission statement.

OBJ 1.10   Is it clear what issues are being promoted?

OBJ 1.11   Are the organization's or person's views on the issues clearly stated?

OBJ 1.12   Is there a clear labeling of editorial and opinion material?

OBJ 1.13   Is there a clear distinction between expressions of opinion on a topic and any informational content that is intended to be objective?

OBJ 1.14   If there is both information-oriented and entertainment-oriented content on the page, is there a clear differentiation between the two?

OBJ 1.15   If there is both advertising and entertainment-oriented content on the page, is there a clear differentiation between the two?

For pages that have a nonprofit or corporate sponsor:

OBJ 1.16   Are the names of any nonprofit or corporate sponsors clearly listed?

OBJ 1.17   Are links included to the sites of any nonprofit or corporate sponsors so that a user may find out more information about them?

OBJ 1.18   Is additional information provided about the nature of the sponsorship, such as what type it is (nonrestrictive, educational, etc.)

CURRENCY (CUR)

Definition: The extent to which material can be identified as up to date.

Questions to Ask

CUR 1.1   Is the date the material was first created in any format included on the page?

CUR 1.2   Is the date the material was first placed on the server included on the page?

CUR 1.3   If the contents of the page have been revised, is the date (and time, if appropriate) the material was last revised included on the page?

CUR 1.4   To avoid confusion, are all dates in an internationally recognized format? Examples of dates in international format (dd mm yy) are 5 June 1997 and 21 January 1999.

CUR 1.5   If the page includes time-sensitive information, is the frequency of updates described?

CUR 1.6   If the page includes statistical data, is the date the statistics were collected clearly indicated, or is there a link to the original data?

CUR 1.7   If the same information also appears in print, such as an online dictionary with a print counterpart, is it clear which print edition the information is taken from? Are the title, author, publisher, and date of the print source listed?

CUR 1.8   If the material was originally presented in broadcast form, is there a clear indication of the date and time the material was originally broadcast?

COVERAGE AND INTENDED AUDIENCE (COV/IA)

Definition of coverage: The range of topics included in a work and the depth to which those topics are addressed.

Questions to Ask

COV/IA 1.1   Is it clear what materials are included at the site?

COV/IA 1.2   If the page is still under construction, is the expected date of completion indicated?

COV/IA 1.3   Is there a print equivalent to the Web page or site? If so, is it clear if the entire work is available on the Web or if parts have been omitted?

COV/IA 1.4   If there is a print equivalent to the Web page, is it clear if the Web version includes additional information not contained in the print version?

COV/IA 1.5   If the material is from a work that is out of copyright (as is often the case with a dictionary or thesaurus) is it clear whether and to what extent the material has been updated?

COV/IA 1.6   If a page incorporates elements of more than one type of page, is there a clear differentiation between the different types of content?

COV/IA 1.7   Is there an adequately detailed description for the products and services offered?

COV/IA 1.8   If the page complements a broadcast or print equivalent to the Web page (i.e., a television show, movie, radio station, etc.) is there an indication of how the broadcast or print equivalent can be accessed?

Definition of intended audience: The group of people for whom material was created.

Questions to Ask

COV/IA 2.1   Is the intended audience for the material clear?

COV/IA 2.2   If material is presented for several different audiences, is the intended audience for each type of material clear?

INTERACTION AND TRANSACTION FEATURES (INT/TRA)

Definition: Interaction features are mechanisms available at a Web site that enable a user to interact with a person or organization responsible for the site. Transaction features are tools that enable a user to enter into a transaction, usually financial via the site.

Questions to Ask

INT/TRA 1.1   If any financial transactions occur at the site, does the site indicate what measures have been taken to ensure their security?

INT/TRA 1.2   If the business, organization, or person responsible for the page is requesting information from the user, is there a clear indication of how that information will be used?

INT/TRA 1.3   If cookies are used at the site, is the user notified? Is there an indication of what the cookies are used for and how long they last?

INT/TRA 1.4   For sites with a membership option, is there a mechanism provided for users to become a member of the organization?

INT/TRA 1.5   Is there a feedback mechanism for users to comment about the site?

INT/TRA 1.6   Is there a mechanism for users to request additional information from the organization or business and if so, is there an indication of when they will receive a response?

INT/TRA 1.7   Are there clear directions for placing an order for items available from the site?

INT/TRA 1.8   Is it clearly indicated when fees are required to access a portion of the site?

INT/TRA 1.9   Are any restrictions regarding downloading and other uses of the materials offered on the page clearly stated?

NAVIGATIONAL AIDS (NAV)

Definition: Elements that help a user locate information at a Web site and allow the user to easily move from page to page within the site. Navigational aids may be text, graphics, or a combination of these.

NAV 1: Browser Titles

Questions to Ask for a Home Page

NAV 1.1   Does the browser title indicate what company, organization, or person is responsible for the contents of the site?

NAV 1.2   Does the browser title indicate that the page is the main, or home page, for the site?

NAV 1.3   Is the browser title short?

NAV 1.4   Is the browser title unique for the site?

Questions to Ask For a Page That Is Not a Home Page

NAV 1.5   Does the browser title indicate what site the page is from?

NAV 1.6   Does the browser title clearly describe the contents of the page?

NAV 1.7   Is the browser title short?

NAV 1.8   Is the browser title unique for the site?

NAV 1.9   Does the browser title reflect the location of the page in the site's hierarchy?

NAV 2: The Page Title

Questions to Ask for a Home Page

NAV 2.1   Does the page title describe what site the page is from? This can be done using a logo.

NAV 2.2   Does the page title indicate that it is the main, or home page, for the site?

NAV 2.3   Is the page title short?

NAV 2.4   Is the page title unique for the site?

Questions to Ask for a Page That Is Not a Home Page

NAV 2.5   Does the page title clearly describe the contents of the page?

NAV 2.6   Is the page title short?

NAV 2.7   Is the page title unique for the site?

NAV 2.8   Does the page title give an indication of the company, organization, or person responsible for the contents of the site?

NAV 3: Hypertext Links

Questions to Ask

NAV 3.1   Does the page include a link to the home page?

NAV 3.2   Does the page include a link to a site map, index, or table of contents?

NAV 3.3   For sites arranged in a hierarchy, does the page include a link to the page one level up in the hierarchy?

NAV 3.4   Are internal directional links consistently placed on each page?

NAV 3.5   For links that access documents at an external site, is there an indication that the user will be leaving the site?

NAV 4: The URL for the Page

Questions to Ask

NAV 4.1   Does the page's URL appear in the body of the page?

NAV 5: The Site Map or Index

Questions to Ask

NAV 5.1   Is there a site map or index on the home page or on a page directly linked from the home page?

NAV 5.2   Does the site map or index include at a minimum the main topics at the site?

NAV 5.3   Is the site map or index easy to read?

NAV 5.4   Is the site map or index organized in a logical manner?

NAV 5.5   Are site map and index entries hypertext links to the referenced material?

NAV 6: Internal Search Engine

Questions to Ask

NAV 6.1   If the site provides large amounts of information, does it include an internal search engine to enable users to locate the information quickly and easily?

NAV 6.2   Does the internal search engine retrieve complete and appropriate results?

NONTEXT FEATURES (NONTX)

Definition: Nontext features are a wide array of elements that require the user to have additional software or a specific browser to utilize the contents of the page. Some examples include image maps, sound, video, and graphics.

Questions to Ask

NONTX 1.1   If the page includes a graphic such as a logo or an image map, is there a text alternative for those viewing the page in text-only mode?

NONTX 1.2   If the page includes a nontext file (such as a sound or video file) that may require additional software, is there an indication of the additional software needed and where it can be obtained?

NONTX 1.3   If a file requires additional software to access it, wherever possible is the same information provided in another format that does not require the additional software?

NONTX 1.4   If a page requires a specific browser or a specific version of a browser does the page specify what is needed and indicate where it can be obtained?

NONTX 1.5   When following a link results in the loading of a large graphic, sound, or video file, is information provided to alert the user that this will happen?

Glossary

Accuracy: The extent to which information is reliable and free from errors.

Advertising: The conveyance of persuasive information, frequently be paid announcements and other notices, about products, services, or ideas.

Advertorial: "An advertisement that has the appearance of a news article or editorial in a print publication" (Richards, 1995–1996, http://advertising.utexas.edu/research/terms/index.html).

Advocacy advertising: "Advertising used to promote a position on a political, controversal or other social issue" (Richards, 1995–1996; http://advertising.utexas.edu/research/terms/index.html).

Advocacy Web page: A page with the primary purpose of influencing public opinion.

Authority: The extent to which material is the creation of a person or organization that is recognized as having definitive knowledge of a given subject area.

Bookmark: A URL address stored on a user's computer that allows the user to easily return to a frequently used page. The ability to store bookmarks is a common browser capability.

Browser: Software on a user's computer that permits both the viewing of and navigation among pages on the World Wide Web.

Browser title: The title of a Web page that is picked up by the browser from the HTML <Title> tag. It usually appears as part of the browser frame at the top of the browser window.

Business Web page: A Web page with the primary purpose of promoting or selling products or services.

Chinese Wall: A policy of separation between the advertising and editorial departments (i.e., the department that produces the informational content) at a publication.

Commercial advertising: "Advertising that involves commercial interests rather than advocating a social or political cause" (Richards, 1995–1996; http://advertising.utexas.edu/research/terms/index.html). It is designed to sell a specific product or service.

Cookies: Data stored by a Web server on a user's computer. This stored information can be read by the Web server when the user returns to the same site. Cookies enable a business to create a shopping cart into which a person can place items to be purchased, and they also allow a site to tailor Web pages to an individual user's preferences.

Copyright: "The protection of the works of artists and authors giving them the exclusive right to publish their works or determine who may so publish" (Gifis, 1996, p. 108).

Corporate sponsor: A business that gives financial or other types of support to something, usually in return for public recognition.

Coverage: The range of topics included in a work and the depth to which those topics are addressed.

Currency: The extent to which material can be identified as up to date.

Date last revised: The date material presented on a Web page was last updated.

Date of creation: The date material presented on a Web page was first created in any format.

Date placed on server: The date material presented on a Web page was first placed on the server.

Entertainment Web page: A page with the primary purpose of providing enjoyment to its users by means of humor, games, music, drama, or other similar types of activities.

Frames: A Web feature that allows the division of a user's browser window into several regions, each of which contains a different HTML page. The boundaries between frames may be visible or invisible. Sometimes each frame can be changed individually, and sometimes one frame in the browser window remains constant while the other frames can be changed by the user.

Graphics: Diagrams, drawings, images, and other types of nontextual material that appear on a Web page.

Home page: The page at a Web site that serves as the starting point from which other pages at the site can be accessed. A home page serves a function similar to the table of contents of a book.

HTML (Hypertext Markup Language): A set of codes that are used to create a Web page. The codes control the structure and appearance of the page when it is viewed by a Web browser. They are also used to create hypertext links to other pages.

Hypertext link (usually referred to just as a link): A region of a Web page that, once selected, causes a different Web page or a different part of the same Web page to be displayed. A link can consist of a word or phrase of text, or an image. The inclusion of hypertext links on a Web page allows users to move easily from one Web page to another.

Index: A listing, often alphabetical, of the major components of a Web site.

Infomercial: "A commercial that is very similar in appearance to a news program, talk show, or other non-advertising program content. The broadcast equivalent of an Advertorial" (Richards, 1995–1996, http://advertising.utexas.edu/research/terms/index.html).

Informational Web page: A page with the primary purpose of providing factual information.

Institutional advertising: "Advertising to promote an institution or organization rather than a product or service, in order to create public support and goodwill" (Richards, 1995–1996; http://advertising.utexas.edu/research/terms/index.html).

Intended audience: The group of people for whom material was created.

Interaction and transaction features: Interaction features are mechanisms available at a web site that enable a user to interact with the person or organization responsible for the site. Transaction features are tools that enable a user to enter into a transaction, usually financial, via the site.

Internal search engine: A search engine that searches for words or phrases only within one World Wide Web site.

JavaScript: A scripting language (a relatively simple computer programming language) that can be embedded in the HTML coding of a Web page. JavaScript can be used to cause the text on a page to change when a mouse is moved across it.

Logo: "A unique trademark, name, symbol, signature, or device to identify a company or other organization" (Weiner, 1996, p. 348).

Meta tags: A group of HTML tags that describe the contents of a Web page. Meta tags do not have to be included on a Web page, and they do not change how the page looks to a user. However, including meta tags on a Web page allows a Web page author to have a certain degree of control over how some search engines index the page.

Navigational aids: Elements that help a user locate information at a Web site and allow the user to easily move from page to page within the site. Navigational aids may be text, graphics, or a combination of these.

News Web page: A page with the primary purpose of providing current information on local, regional, national, or international events, or providing current information about a particular topic, such as business news, legal news, and so forth.

Nonprofit sponsor: An individual or nonprofit organization that provides financial or other types of support for something, usually in return for public recognition.

Nontext features: A wide array of elements that require a user to have additional software or a specific browser to utilize the contents of a Web page. Some examples of nontext features include graphics, image maps, sound, and video.

Objectivity: The extent to which material expresses facts or information without distortion by personal feelings or other biases.

Page title: The title found in the text of the Web page (as distinguished from the browser title that usually appears at the very top of the screen).

Personal Web page: A page created by an individual who may or may not be affiliated with a larger institution. Personal pages often showcase an individual's artistic talents or are devoted to a favorite hobby or pastime.

Search Engine: A tool that can search for words or phrases on a large number of World Wide Web pages. Examples of search engines include AltaVista, Infoseek, and HotBot.

Secure transaction: An encrypted communication between a Web server and a Web browser. Because the data communicated in a secure transaction are encrypted, or scrambled, the opportunity for the content to be read by an unauthorized person during the transfer across the Web is minimized. Financial transactions conducted over the World Wide Web are frequently made as secure transactions.

Site map: A display, often graphical, of the major components of a Web site.

Sponsorship: Financial or other support given by an individual, business, or organization for something, usually in return for some form of public recognition. (See also Corporate sponsor and Nonprofit sponsor.)

Uniform Resource Locator (URL): A unique identifier that distinguishes a Web page from all other World Wide Web pages.

Web page: An HTML file that has a unique URL address on the World Wide Web.

Web site: A collection of related Web pages interconnected by hypertext links. Each Web site usually has a home page that provides a table of contents to the rest of the pages at the site.

Web subsite: A site on the World Wide Web that is nested within the larger Web site of a parent organization. The parent organization often has publishing responsibility for the subsite, and the URL for the subsite is usually based on the parent site's URL.

Word of mouth advertising: The endorsement of a product or service by an individual who has no affiliation with that product or service other than being a user of it, and who is not paid for the endorsement.

References

Ad Council. (1997). [Ad Council Web site]. [Online]. Available: http://www.adcouncil.org.

American Society of Magazine Editors. (1997, October 3). ASME guidelines for new media [Online]. Available: http://webreview.com/97/10/03/feature/guide.html.

Copyright Act of 1976, 17 U.S.C. Sect. 101 et seg. Available: Lexus-Nexis Academic Universe.

Gifis, S. H. (1996). *Law dictionary* (4th ed.). Hauppauge, NY: Barron's

Rankin, B. (1998, June 30). Best of Tourbus #3: An even closer look at cookies. *The Internet Tourbus* [Online]. Available: http://www.tourbus.com/archive/tb063098.htm .

Richards, J. (1995–1996). *Dictionary of terminology, advertising* [Online]. Available: http://advertising.utexas.edu/research/terms/index.html.

Stout, D. (1998, June 6). Oops. Bob Hope is not dead. *The New York Times*, p. A9.

Weiner, R. (1996). Webster's new world dictionary of media and communications (rev. ed.). New York: Macmillan.

Bibliography

Ad Council. (1997). [Ad Council Web site]. [Online]. Available: http://www.adcouncil.org [November 1997–May 1998].

AdMedium Newsletter. (1998, January). [Online]. Available: http://uts.cc.utexas.edu/~admedium/newsletter.html [1998, February 17].

Alexander, J., & Tate, M. A. (1996–1998). *Evaluating Web resources* [Online]. Available: http://www.science.widener.edu/~withers/webeval.htm [1996–1998]

Alexander, J., & Tate, M. A. (1996, November–December). Teaching critical evaluation skills for World Wide Web resources. *Computers in Libraries, 16*(10), 49–55.

Arens, W. F. (1996). *Contemporary advertising* (6th ed.). Chicago: Irwin.

Azcuenaga, M. L. (1994). *Advertising: Interpretation and enforcement policy* [Online]. Remarks given before the American Advertising Federation, 1994 National Government Affairs Conference, Washington, DC. Available: http://www.ftc.gov/speeches/azcuenaga/aaf94.htm [1998, February 17].

American Society of Magazine Editors. (1997, October 3). *ASME guidelines for new media* [Online]. Available: http://webreview.com/97/10/03/feature/guide.html [1998, February 26].

Andrews, W. (1996, June 3). Sites dip into cookies to track user info. *WebWeek, 2*(7), pp. 17, 20.

Barnouw, E. (1989). *International encyclopedia of communications.* New York: Oxford University Press.

Benjamin, B. (1996, August 29). *Elements of Web design* [Online]. Available: http://www.cnet.com/Content/Features/Howto/Design/?dd [1997, July 16].

Berger, J. (1995, December 30). Critics of traditional media are flocking to the Web. *Editor & Publisher,* [Online], *128*(52), 2 pages. Available: Periodical Abstracts, [1997].

Berners-Lee, T. (1995, May). *Style guide for online hypertext* [Online]. Available: http://www.w3.org/pub/WWW/Provider/Style/All.html [1997, June 5].

Beyer, B. K. (1995). *Critical thinking.* Bloomington, IN: Phi Delta Kappa Educational Foundation.

Blackwood, F., Moore, J., & Yee, B. (1997, February). Compete using technology. *Home Office Computing, 15*(2), 60–66.

Booker, E. (1996, April 1). What's your URL, doc? Health care sites offer cheap way to reach patients. *WebWeek, 2*(4), p. 20.

Bort, J. (1996, September 2). The key to security. *InfoWorld, 18*(36), pp. 1, 51–52.

Borzo, G. (1995, July 31). E-publish or perish? *American Medical News* [Online], *38*(28). Available: Proquest Direct [1998, March 27].

Brandt, D. S. (1996, May). Evaluating information on the Internet. *Computers in Libraries, 16*(5), 44–46.

Brinkley, M., & Burke, M. (1995). Information retrieval from the Internet: An evaluation of the tools. *Internet Research: Electronic Networking Applications and Policy, 5*(3), 3–10.

Britt, S. H. (1994). Advertising. In *The Encyclopedia Americana* (Intl. Ed., Vol. 1, pp. 195–206). Danbury, CT: Grolier.

Burgstahler, S., Comden, D., & Fraser, B. M. (1997). Universal access: Designing and evaluating Web sites for accessibility. *Choice, 34* (Suppl.), 19–22.

Choice: Current [Web] reviews for academic libraries. (1998). *Choice, 35* (Suppl.).

Ciolek, T. M. & Goltz, I. M. (Eds.). (1995–1998). *Information Quality WWW Virtual Library: The Internet guide to construction of quality online resources.* [Online]. Available: http://www.ciolek.com/WWWVL-InfoQuality.html [1996, April–1998, October].

Ciolek, T. M., & Goltz, I. M. (1997–1998). *WWW.CIOLEK.COM: Asia Pacific Research Online* [Online]. Available: http://www.ciolek.com/home.html [1997, September–1998, October].

Clark, S. (1997–1998). Back to basics: META tags. *Webdeveloper.com* [Online].Available: http://www.webdeveloper.com/categories/html/html_metatags.html.

Cohen, L. B. (1997). Librarians on the Internet: The search for quality begins. *Choice, 34* (Suppl.), 5–17.

Cohen, L. B. (1998). Searching for quality on the internet: Tools and strategies. *Choice, 35* (Suppl.), 11–27.

Consumer group certifies 1,000 businesses. (1996, November 18). *WebWeek*, p. 26.

Consumers Union of U.S. (1998) [Consumer Reports Web site]. [Online]. Available: http://www.consumerreports.com [1998, March].

Cookie Central. (1997–1998). *Cookies* [Online]. Available: http://www.cookiecentral.com/cm002.htm [1998, March 10].

Cookie Central. (1997–1998). *Persistent cookie FAQ* [Online]. Available: http://www.cookiecentral.com/faq.htm [1998, March 10].

Copyright Act of 1976, 17 U.S.C. Sect. 101 et seq. Available: Lexis–Nexis Academic Universe [1998, October 7].

Corel, Gallery 2. [Computer software]. (1995). Ottawa, Ontario, Canada: Corel Corporation.

Cottrell, J., & Eisenberg, M. B. (1997, May). Web design for information problem-solving: Maximizing value for users. When you design your web pages, are you really keeping your potential users in mind? *Computers in Libraries, 17*(5), 52–57.

Council of Better Business Bureaus. (1997–1998). [Better Business Bureau Web site]. Available: http://www.bbb.org [1998, March 26].

Council of Better Business Bureaus. (1996, March 25). *Better Business Bureau urges businesses to adhere to ethical practices in the online marketplace* [Online]. Available: http://www.bbb.org/alerts/rel-onln.html [1996, June 14].

Crigger, B. (1995, September). So now we are asking the world. *Hastings Center Report* [Online], *25*(5), 47. Available: Periodical Abstracts [1996, June 6].

Criner, K. & Wilson, J. (1996, August 31). Big lessons in big Web sites. *Editor & Publisher* [Online], *129*(35), pp. 6, 34. Available: Periodical Abstracts [1997].

Csatari, J. (1996, January). Hacker quacks. *Men's Health* [Online], *11*(1), 17. Available: Periodical Abstracts [1996, June 4].

Davey, T. (1996, October 28). Making the Web safe for commerce. *PC Week*, p. 44.

Donnelly, David F. (1996). Selling on, not out, the Internet. *Journal of Computer-Mediated Communication* [Online], *2*(1). Available: http://www.asusc.org/jcmc/vol2/issue1/adsnew.html [1998, February 17].

Doyle, B., Modahl, M. A., & Abbot, B. (1997, May 5). What advertising works. *Brandweek* [Online], (Why the Web? The Case for Internet Advertising Suppl.), 38–42. Available: Proquest Direct [1997, December 13].

Duval, B. K. & Main, L. (1996, Fall). Building Web pages: An update. *Library Software Review* [Online], *15*(3), 158–162. Available: Periodical Abstracts [1997].

Eichelberger, L. (1997). *The Cookie controversy* [Online]. Available: http://www.cookiecentral.com/ccstory/index.html [1998, March 17].

Elkin-Koren, N. (1996, September). Public/private and copyright reform in cyberspace. *Journal of Computer-Mediated Communication* [Online], *2*(2). Available: http://www.ascusc.org/jcmc/vol2/issue2/elkin.html [1998, February 17].

Fay, S. M. (1997). Cyberspace: Stretching the fabric of copyright law. *Ohio Northern Law Review* [Online], *23*, 15739 words. Available: Lexis–Nexis Universe [1997, November 11].

Feldman, J. (1996, October 1). Advertising and the law in cyberspace. *Folio* [Online], *25*(14), 47–48. Available: Periodical Abstracts [1997].

Felser, A. J. (1996, January). Are your ads illegal? *Home-Office Computing* [Online], *14*(1), 92-93. Available: Periodical Abstracts [1997].

Fishman, S. (1997). *The copyright handbook: How to protect & use written works* (4th ed.). Berkeley, CA: Nolo Press.

Fitzgerald, M. (1995, February 11). AP chief: Beware of yellow journalism in cyberspace. *Editor & Publisher* [Online], *128*(6), 31. Available: Periodical Abstracts [1997].

Forcht, K. A., & Fore, R. E., III. (1995). Security issues and concerns with the Internet. *Internet Research: Electronic Networking Applications and Policy 5*(3), 23–31.

Foster, E. (1996, July 22). Can mixing 'cookies' with online marketing be a recipe for heartburn? *InfoWorld, 18*(30), 54.

Frappaolo, C. (1996, September). Control your documents via the Internet. *Databased Adviser, 14*(9), 46–50.

Fraser, B., Comden, D. & Burgstahler, S. (1998). Including users with disabilities: Designing library web sites for accessability. *Choice, 35* (Suppl.), 35–37.

Gardner, E. (1996, November 4). Business grows for site-testing labs. *WebWeek, 2*(17), p. 61.

Gibeaut, J. (1997, February). Intellectual property law. *ABA Journal* [Online], *83*, 2544 words. Available: Lexis–Nexis Universe [1997, October 30].

Gifis, S. H. (1996). *Law dictionary* (4th ed.). Hauppauge, NY: Barron's.

Grassian, E. (1995–1998). Thinking critically about World Wide Web resources [Online]. Available: http://www.library.ucla.edu/libraries/college/instruct/web/critical.htm [1998, March 24–1998, October].

Guernsey, L. (1996, January 12). Cyberspace citations. *Chronicle of Higher Education* [Online], A18–A21. Available: Periodical Abstracts [1997].

Haley, E. (1996, June). Exploring the construct of organization as source: Consumers' understandings of organizational sponsorship of advocacy advertising. *Journal of Advertising* [Online], *25*(2). Available: Lexis–Nexis Universe [1998, February 26].

Hall, P. (1997, May–June). Log on tomorrow. *Print* [Online], *51*(3). Available: Infotrac Searchbank/Expanded Academic Index/Article A19672249 [1997, July 25].

Harris, R. (1997). *Evaluating Internet research sources* [Online]. Available: http://www.sccu.edu/faculty/R_Harris/evalu8it.html [1998, March 31].

Hatlestad, L. (1996, September 9). The encryption prescription. *InfoWorld, 18*(37), pp. 1, 16.

Hauptman, R., & Motin, S. (1994, March). The inverted file: The Internet, cyberethics, and virtual morality [Guest editorial]. *Online, 18*(2), 8–9.

Hernon, P. (1995). Discussion forum: Disinformation and misinformation through the Internet: Findings of an exploratory study. *Government Information Quarterly, 12*(2), 133–139.

Hinchliffe, L. J. (1994–1997). *Resource selection and information evaluation* [Online]. Available: http://alexia.lis.uiuc.edu/~janicke/Evaluate.html [1998, March 24].

Ho, J. (1997, June). Evaluating the World Wide Web: A global study of commercial sites. *Journal of Computer Mediated Communication* [Online], *3*(1), 32 printed pages. Available: http://www.ascusc.org/jcmc/vol3/issue1/ho.html [1998, February 17].

Hockey, S. (1994, Spring). Evaluating electronic texts in the humanities. *Library Trends, 42*(4), 676–693.

Hoggart, R. (1995, May 5). The uses of computeracy. *New Statesman & Society* [Online]. *8*(351), 21–23. Available: Periodical Abstracts/UMIACH16700.00 [1996, June 6].

Horton, R. (1997, May 17). Sponsorship, authorship, and a tale of two media. *Lancet* [Online], *349*(9063), 1411–1412. Available: Periodical Abstracts [1997].

How to read a page: Multiple ways of reading Web pages and sites [Online]. (1996–1997). Available: http://cal.bemidji.msus.edu/English/Morgan/Courses/EN293/HowToReadPage.html [1997, May 15].

Hunt, K. (1996, November). Establishing a presence on the World Wide Web: A rhetorical approach. *Technical Communication, 43*(4), 376–387.

Infoseek. (1995–1997). *Additional information on submitting a Web site* [Online]. Available: http://www.infoseek.com/Help?sv=IS&lk=noframes&pg=meta_tag.html [1997, October 30].

Jade River Designs. (1996). *The 6 myths of Web marketing and what you really need to know* [Online]. Available: http://www.jaderiver.com/webmyth.html [1997, July 16].

Jade River Designs. (1996). *Paper brochure vs. online brochure* [Online]. Available: http://www.jaderiver.com/paperweb.html [1997, July 16].

Jones, D. (1996). *Critical thinking in an online world* [Online]. Available: http://www.library.ucab.edu/untangle/jones.html [1996, October 18].

King, B. (1996). Web of deceit. *Internet Underground, 7*, 26–32.

Klobas, J. E. (1995). Beyond information quality: Fitness for purpose and electronic information resource use. *Journal of Information Science, 21*(2), 95–114.

Koschnick, W. J. (1995). *Dictionary of marketing.* Brookfield, VT: Gower.

Lands' End Direct Merchants. (1998). *Welcome to Lands' End* [Online]. Available: http://www.landsend.com [1998, February–March].

Levine, R. (1995). *Guide to Web style* [Online]. Available: http://www.sun.com/styleguide/tables/Printing_Version.html [1997, June 5].

Libraries of Purdue University. (1997). *Anyone can (and probably will) put anything up on the Internet* [Online]. Available: http://thorplus.lib.purdue.edu/~techman/eval.html [1998, March 4].

Library of Congress. Copyright Office. (1997–1998). U.S. Copyright Office home page. [Online]. Washington, DC: Library of Congress, Copyright Office. Available: http://lcweb.loc.gov/copyright [1998, March–1998, October].

Makulowich, J. S. (1996, February–March). Quality control on the Net. *Database, 19*(1), 93–94.

Math Forum. [The Math Forum Web Site]. [Online]. Available: http://www.forum.swarthmore.edu [1998, March–June].

McCandlish, S. (1995, December 26). *How to make your Web offerings useful to the most visitors* [Online]. Available: http://www.eff.org/~mech/Scritti/html.tipsheet [1997, July 20].

McClain, J. D. (1997, December 16). Children's privacy violated on Net, FTC says. *The Philadelphia Inquirer*, pp. C1, C8.

Minnesota Public Radio. (1998). [*MPR, Minnesota Public Radio Web site*]. [Online]. Available: http://www.mpr.org [1998, March–1998, October].

Minnesota Public Radio. (1998). *A prairie home companion, from Minnesota Public Radio* [Online]. Available: http://phc.mpr.org [1998, March–1998, October].

Monnet, B. J. (1995). *The quality of electronic information products and services* [Online]. Available: http://www2.echo.lu/impact/imo/9504.html#criteria [1996, August 8].

Moran, S. (1996, August 5). Better Business Bureau targets electronic fraud. *WebWeek*, p. 19.

Morris, M., & Ogan, C. (1996, Winter). The Internet as mass medium. *Journal of Communication* [Online], *46*(1), 39–50. Available: Periodical Abstracts [1996, June 6].

Mullin, E. (1996, November 4). Using the META tags to refresh pages and redirect visitors. *Web Week*, pp. 61–62.

Murphy, K. (1996, June 17). *Debate over advertising standards. WebWeek*, p. 24.

Murphy, K. (1996, February 1). Sending legal documents overseas can be tricky, but 'Cybernotaries' may help. *WebWeek*, p. 7.

Narayan, S. (1996, October 21). Cookies give users automatic log-in option: A convenience, but also an intrusion? *WebWeek, 2*(16), pp. 51–52.

National Center for Supercomputing Applications (NCSA). (1997, June 19). *Multimedia design for the World Wide Web* [Online]. Available: http://www.ncsa.uiuc.edu/General/Training/AdvHTML/multimedia.design.html

National Public Radio. (1998). *NPR online, National Public Radio* [Online]. Available: *http://www.npr.org* [1998, March 27].

Netscape Communications Corporation. (1998). *Cookies and privacy FAQ* [Online]. Available: http://search.netscape.com/assist/security/faqs/cookies.html [1998, March 18].

Newhagen, J. E., & Rafaeli, S. (1996, Winter). Why communication researchers should study the Internet. *Journal of Communication* [Online], *46*(1), 4–13. Available: Periodical Abstracts [1996, June 6].

Notess, G. R. (1996, February–March). News resources on the World Wide Web. *Database, 19*(1), 12–20.

O'Harrow, R., Jr. (1998, March 9). Picking up on 'cookie' crumbs: Web sites want to track your every move. Should you let them? *The Washington Post* [Online], p. F25 (790 words). Available: Lexis–Nexis Universe [1998, March 10].

Ormondroyd, J., Engle, M., & Cosgrave, T. (1995–1996). *How to critically analyze information sources* [Online]. Available: http://www.library.cornell.edu/okuref/research/skill26.htm [1998, March 24].

Ostrow, R., & Sweetman, R. S. (1988). *The dictionary of marketing*. New York: Fairchild.

Pagell, R. A. (1995, July–August). Quality and the Internet: An open letter [Guest editorial]. *Online*, 7–9.

Pfaffenberger, B. (1996). *Web search strategies*. New York: MIS Press.

Physicians for Social Responsibility (PSR). (1998). [PSR, Physicians for Social Responsibility Web site]. [Online]. Available: http://www.psr.org [1998, February–March].

Public Broadcasting Service (PBS). (1998). [Public Broadcasting Service Web site]. [Online]. Available: http://www.pbs.org [1998, March].

Rankin, B. (1998, June 30). Best of Tourbus #3: An even closer look at cookies. *The Internet Tourbus* [Online]. Available: http://www.tourbus.com/archive/tb063098.htm [1998, August 20].

Reichertz, P. S. (1997). Understanding government regulation of the marketing and advertising of medical devices, drugs, and biologics: The challenges of the Internet. *Food and Drug Law Journal 52*(33), 3533 words. Available: Lexis–Nexis Universe [1997, November 9].

Richards, J. (1995–1996). Dictionary of terminology, advertising [Online]. Available: http://advertising.utexas.edu/research/terms/index.html [1998, February–March].

Robinson, K. (1997–1998). *Literature by Ken* [Online]. Available: http://www.personal.psu.edu/staff/k/j/kjr106/lit.html [1997, December–1998, March].

Robischon, N. (1998, July/August). Browser beware: As search sites battle for profit, they're not telling us what is an ad and what isn't. *Brill's Content, 1*(1), 4–44.

Rosenberg, J. M. (1995). *Dictionary of marketing and advertising*. New York: Wiley.

Rothstein, L. (1997, May–June). Happy birthday, Mr. Lodestar. *Bulletin of the Atomic Scientists* [Online], *53*(3). Available: Infotrac Searchbank/Expanded Academic Index ASAP/Article A19368174 [1997, July 25].

Ruiz, F. (1997, November 10). Getting rid of cookies crumbs free. *The Tampa Tribune* [Online], p. 4 (562 words). Available: Lexis–Nexis Universe [1998, March 10].

Scholz, A. (1996). *Evaluating World Wide Web information* [Online]. Available: http://thorplus.lib.purdue.edu/research/classes/gs175/3gs175/evaluation.html [1997].

Schrock, K. (Ed.). (1995–1998). *Kathy Schrock's guide for educators: Critical evaluation surveys* [Online]. Available: http://www.capecod.net/schrockguide/eval.htm [1998, March 24].

Search engines crack down on META tag "abuse." (1997, November 10). *Internet Business Advantage* [Online], 2 printed pages. Available: http://www.cobb.com/iba/9711a/iba97ba4.htm [1997].

Silverblatt, A. (1995). *Media literacy: Keys to interpreting media messages*. Westport, CT: Greenwood.

Smithsonian Institution. (1998). [Smithsonian Institution Web site]. [Online]. Available: http://www.si.edu [1998, March 21].

Snyder, J. M. (1995, August). Research on the Internet: Inside risks. *Communications of the ACM, 38*(8),130.

Starek, R. B. (1997, June 25). *Protecting the consumer in the global marketplace* [Online]. Remarks given at the Agenda for Action: New Perspectives in Consumer Affairs and Consumer Markets, Institute of Trading Standards Administration, Antwerp, Belgium. Available: *http://www.ftc.gov/speeches/starek/antwefin.htm* [1998, February 17].

Stefanac, S. (1996, June). Copyright ain't dead yet. *Macworld, 13*(6), 137–138.

Stein, L. D. (1997). *How to set up and maintain a Web site* (2nd ed.). Reading, MA: Addison-Wesley.

Stout, D. (1998, June 6). Oops. Bob Hope is not dead. *The New York Times*, p. A9.

Sullivan, E. (1996, June 24). Are Web-based cookies a treat or a recipe for trouble? Protocol makes it easier to implement web-based applications, but users should be careful about what they store. *PC Week*, pp. 75, 91.

Taylor, D. (1996, November 11). Make your intranet international, just like your company. *InfoWorld, 18*, pp. 46, IW12.

Tillman, H. N. (1995–1998). *Evaluating quality on the net* [November 8, 1997 version]. [Online]. Available: http://www.tiac.net/users/hope/findqual.html [1998, March 24].

Treloar, A. (1995–1997). *Scholarly publishing and the fluid World Wide Web* [Online]. Available: http://www.csu.edu.au/special/conference/apwww95/papers95/atreloar/atreloar.html [1998, March 24].

Trustees of The University of Pennsylvania. (1994–1998). *OncoLink* [Online]. Available: *http://www.oncolink.upenn.edu* [1998, February].

Twitchell, J. B. (1996). *Adcult USA: The triumph of advertising in American culture.* New York: Columbia University Press.

United States Environmental Protection Agency. (1998). [United States Environmental Protection Agency homepage] [Online]. Available: http://www.epa.gov/ [1998, February–March].

United States Office of the President. (1998). *Welcome to the White House* [Online]. Available: http://www.whitehouse.gov [1998, March 27].

University of Northumbria Information Services Department. (1996–1998). *Web search tool features* [Online]. Available: http://www.unn.ac.uk/features.htm [1997–1998, October].

University of Pennsylvania. (1997). *Welcome to the University of Pennsylvania* [Online]. Available: http://www.upenn.edu [1997–1998, October].

University of Texas at Austin Department of Advertising. (1995–1997). *Advertising world* [Online]. Available: http://advertising.utexas.edu/world/index.html [FebruaryMarch 1998].

Uretsky, S. (1996, February). Bad medicine: Beware of useless or dangerous medical advice online. *Internet World, 7*(2), 54–55.

Van Name, M. L., & Catchings, B. (1996, November 11). Habla usted or parlez-vous the Web? *PC Week*, p. N6.

Washington Post Co. (1998). *Washingtonpost.com* [Online]. Available: http://www.washingtonpost.com [1998, February]

Weibel, S. L. (1995, Spring). The World Wide Web and emerging Internet resource discovery standards for scholarly literature. *Library Trends, 43*(4), 627–644.

Weiner, R. (1996). *Webster's new world dictionary of media and communications* (rev. ed.). New York: Macmillan.

Whalen, D. (1998). *The unofficial cookie FAQ.* (Version 2.0) [Online]. Available: http://www.cookiecentral.com/unofficial_cookie_faq.htm [1998, March 10].

Why? InterNetworking. (1998). *Welcome to the White House* [Online]. Available: http://www.whitehouse.net [1998, March].

Williams, R., & Tollett, J. (1998). *The non-designer's Web book: An easy guide to creating, designing, and posting your own Web site.* Berkeley, CA: Peachpit Press.

Wilson, S. (1995). *World Wide Web design guide.* Indianapolis, IN: Hayden Books.

Young, J. R. (1997, August 15). Purported MIT commencement speech turns out to be another Internet hoax. *Chronicle of Higher Education, XLIII*(49), A20.

Ziegler, B. (1996, July 25). Old fashioned ethic of separating ads is lost in cyberspace. *The Wall Street Journal*, p. B1.

Index

A

ACC 1.1
  illustration of as basic element, 43
  illustration of in informational page, 43, 79
  in checklist of basic elements, 56
ACC 1.2
  illustration of as basic element, 43
  illustration of in informational page, 43, 79
  in checklist of basic elements, 56
ACC 1.3
  in informational checklist, 81
ACC 1.4
  illustration of in informational page, 79
  in checklist of basic elements, 56
ACC 1.5
  illustration of as basic element, 43
  illustration of in informational page, 43
  illustration of in news page, 86
  in informational checklist, 81
  in news checklist, 87
ACC 1.6
  in business checklist, 68
Accuracy (see also ACC checklist items)
  condensation of traditional publishing process, 12
  definition, 11, 42
  discussion of, 42
  discussion of in OncoLink screen capture, 42
  illustration of in OncoLink screen capture, 43
  in checklist of basic elements, 56
  of business pages, 68
  of informational pages, 81
  of news pages, 87
  of traditional sources, 11
  of Web sources, 12
  questions to ask, 42
Ad Council Web pages
  discussion of advocacy advertising, 21
  illustration of advocacy advertising, 22
  illustration of consistent design, 104
  mission statement, 21
Advertising
  blending of types, 24

blending with information, 26
blending with information and entertainment, 27
blending with information, continuum, 27
definition, 19
identifying purpose for providing information, 30
types of, 19–24
Advertising and information
  blending of, 26
  identifying key players, 30
  questions to ask, 29
  relation between, 29
  visual distinctions between, 26
Advertising and sponsorship
  blending of, 24
  distinction between, 24
  potential influence on information, 26–34
Advertising policy, illustration of at CIOLEK Web site, 33
Advertising supported sites, strategies for evaluating, 30
Advertising, advocacy (see Advocacy advertising)
Advertising, commercial (see Commercial advertising)
Advertising, information, and entertainment
  blending of, 27
  screen capture illustration of blending, 35
Advertising, institutional (see Institutional advertising)
Advertising, word of mouth (see Word of mouth advertising)
Advertorial, 26
Advocacy advertising
  Ad Council screen capture illustration, 22
  Ad Council Web site, 21
  definition, 21
Advocacy Checklist, 59–62
Advocacy pages
  analysis of, 59
  checklist, 59–62
  definition, 58
  discussion of, 58–62
  keys to recognizing, 58
American Society of Magazine Editors, 26